AF476912

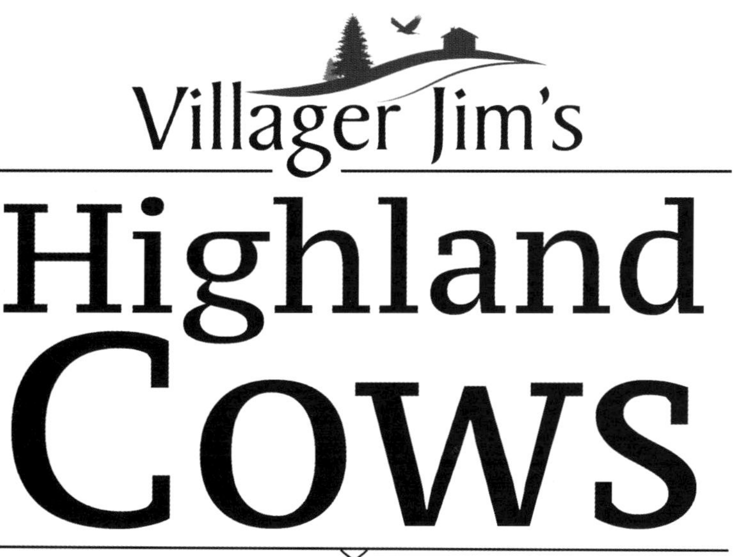

Villager Jim's
Highland
Cows

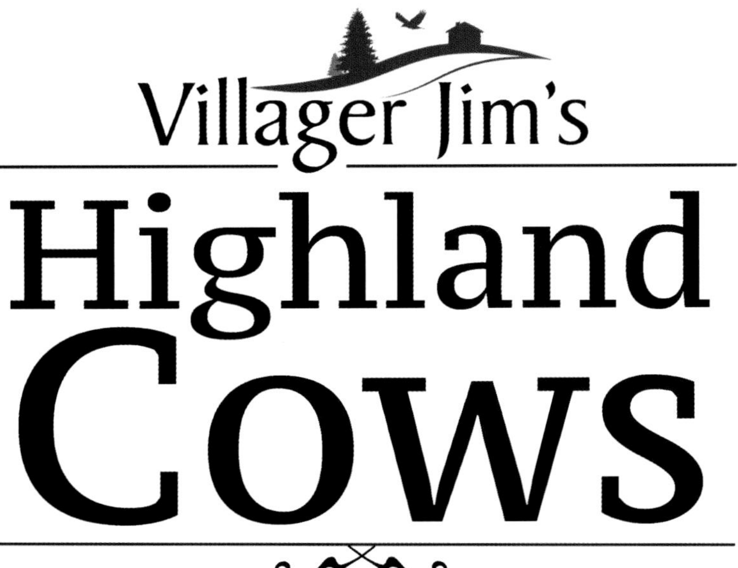

Villager Jim's
Highland Cows

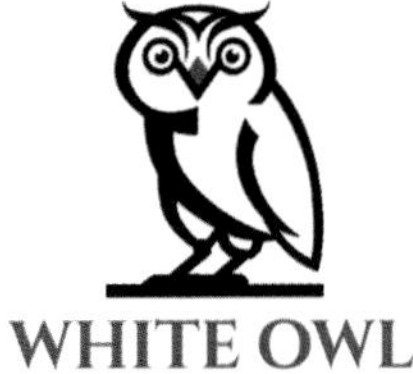

WHITE OWL

First published in Great Britain in 2018 by
PEN & SWORD WHITE OWL
an imprint of
Pen & Sword Books Limited
47 Church Street
Barnsley
South Yorkshire
S70 2AS

Copyright © Villager Jim 2018

ISBN 978 1 5267 0683 6

Printed and bound in India by Replika Press Pvt. Ltd.

Pen & Sword Books Limited incorporates the imprints of Atlas, Archaeology, Aviation, Discovery, Family History, Fiction, History, Maritime, Military, Military Classics, Politics, Select, Transport, True Crime, Air World, Frontline Publishing, Leo Cooper, Remember When, Seaforth Publishing, The Praetorian Press, Wharncliffe Local History, Wharncliffe Transport, Wharncliffe True Crime and White Owl.

For a complete list of Pen & Sword titles please contact:
PEN & SWORD BOOKS LIMITED
47 Church Street, Barnsley, South Yorkshire, S70 2AS, United Kingdom
E-mail: enquiries@pen-and-sword.co.uk
Website: www.pen-and-sword.co.uk

Or
PEN AND SWORD BOOKS
1950 Lawrence Rd, Havertown, PA 19083, USA
E-mail: Uspen-and-sword@casematepublishers.com
Website: www.penandswordbooks.com

Introduction

There is something quietly historic about Highland cattle, as though they were here long before us. They are my favourite farm animal and I am lucky enough to be able to photograph them on the moorlands surrounding my home in the Peak District National Park. For me, it is the gentle giant syndrome; they are great big woolly monsters with fierce horns, yet so docile – unless you get too close to their young ones! Their fantastic hairy coats and eyes hidden deep inside matted dreadlocks give them a mystical appearance.

Highlands wander slowly over whatever ground you give them, be it moorland, heathland, lush green farming fields or woodland copses. The moorlands can be a really tough place in winter, often very wild and windy. Highlands' coats help keep them warm in atrocious conditions, but of course the flip side is that in the long hot summer days, they can get overheated. I often see them standing in the local rivers cooling their hooves, which must seem bloomin' lovely to them in those very hot spells. They will stand for ages in the same place just letting the cooling moorland streams trickle by taking off the worst of the heat.

My business partner Fiona breeds Highlands on her family's smallholding in Rutland, and her father Jeff is president of the Highland Cow Society. They have a herd of about twenty stunning animals that regularly appear in my photographs. When we make a business visit to Fiona and Darren (Fiona's husband) we always take the opportunity to see 'the gang'. Jimbo (great name choice!), the herd leader, is an absolutely ginormous bull. The cows and bulls in the herd are such wonderful creatures and it's obvious for all to see that they are very much loved. Like humans, some Highlands can be soft and gentle and others a little gnarly, but Fiona's herd are the most affable of characters – including Ellie, Boo Boo, Tilly, Maggie May and Belle (and her calf, Baby Belle). Fiona often sits with them and they know she is their friend. I think they treat her as one of the herd. (Sorry partner, but it's true!) They have won all sorts of prizes and stand out in the show ring as being extremely well looked after.

To photograph Highlands, moving slowly, always within their field of view, is the best way of gaining their trust. The golden rule is to make sure you have been noticed; animals don't like being crept up on. I will often walk up to them talking away as I'm convinced that this extra level of intimacy makes them feel less threatened by my presence. Although they are normally calm, when it is calving time, or their calves are young, the cows' strong maternal instinct can put you in danger if you push their patience too far. I would usually advise against going into a field with youngsters unless you are quite a distance away. Give them that extra space needed to make them feel confident that you won't harm their young.

Calves are often so entertaining to watch. They love hanging around with their mums and gathering in a playtime crèche. It is surprising how fast they can run so you have to be on your guard amidst teenagers and younger ones playing in a field. Usually the adults will just do their own thing without any acknowledgement that they have seen you, which is the way I like it so I can roam amongst them taking shots from many different angles.

Some of the best photographs can be achieved by getting down low and shooting at their level or below, which can create a lot of atmosphere in a picture. Also, it is good to just get the nose or the horns, or have the animal only partly in shot looking into the rest of the composition. It is amazing how much fun you can have with them on the edge of a picture as your brain knows they are big, so the rest is left to the imagination. This is what I advised Helen Skelton to do when she and the BBC *Countryfile* team came to film me – although she was photographing Limousins rather than Highlands – and she ended up taking some super pictures. We had the most fantastic day and it was a great honour to have been asked to be part of the programme.

I run photography courses with a friend and we often come across Highlands on our journeys out. They tend to move as a herd very slowly so when you see them at a distance you know they will still be there when you get to the location. This is something you can't say about a lot of other animals that are out and about in the Peaks. Highland cattle make superb compositions, especially on the moorland, which in itself is wonderful to photograph. Having a Highland as the subject with a moorland backdrop really makes a double whammy of a shot!

One of my favourite photographs of a Highland cow is one I've called *Mistic Meg* – with the deliberate misspelling. It was taken one very misty morning when I came across a herd in a small wood. I found one cow looking at me through the mist from behind a tree trunk, looking both mysterious and also very funny. When I put the photograph on my Facebook page, it went crazy, with many thousands of 'likes'.

There aren't many better things to do in life on a summer's day than to sit amongst a herd of Highlands in the heather on the moors and enjoy the beautiful countryside in the presence of these wonderfully majestic creatures.

Villager Jim

Visit Villager Jim's Facebook page at **www.facebook.com/villagerjim**
And his website at **www.villagerjim.com**

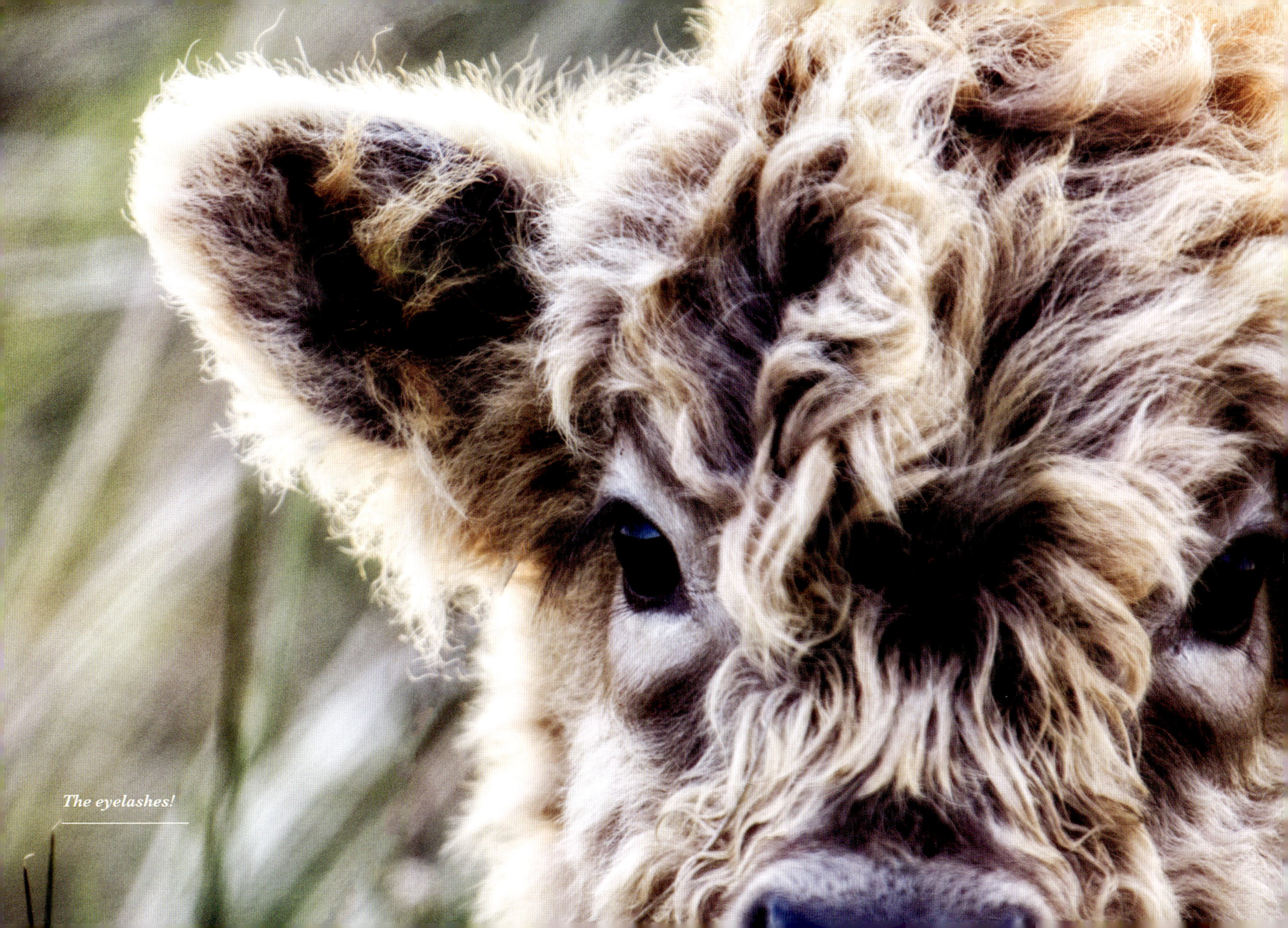

The eyelashes!

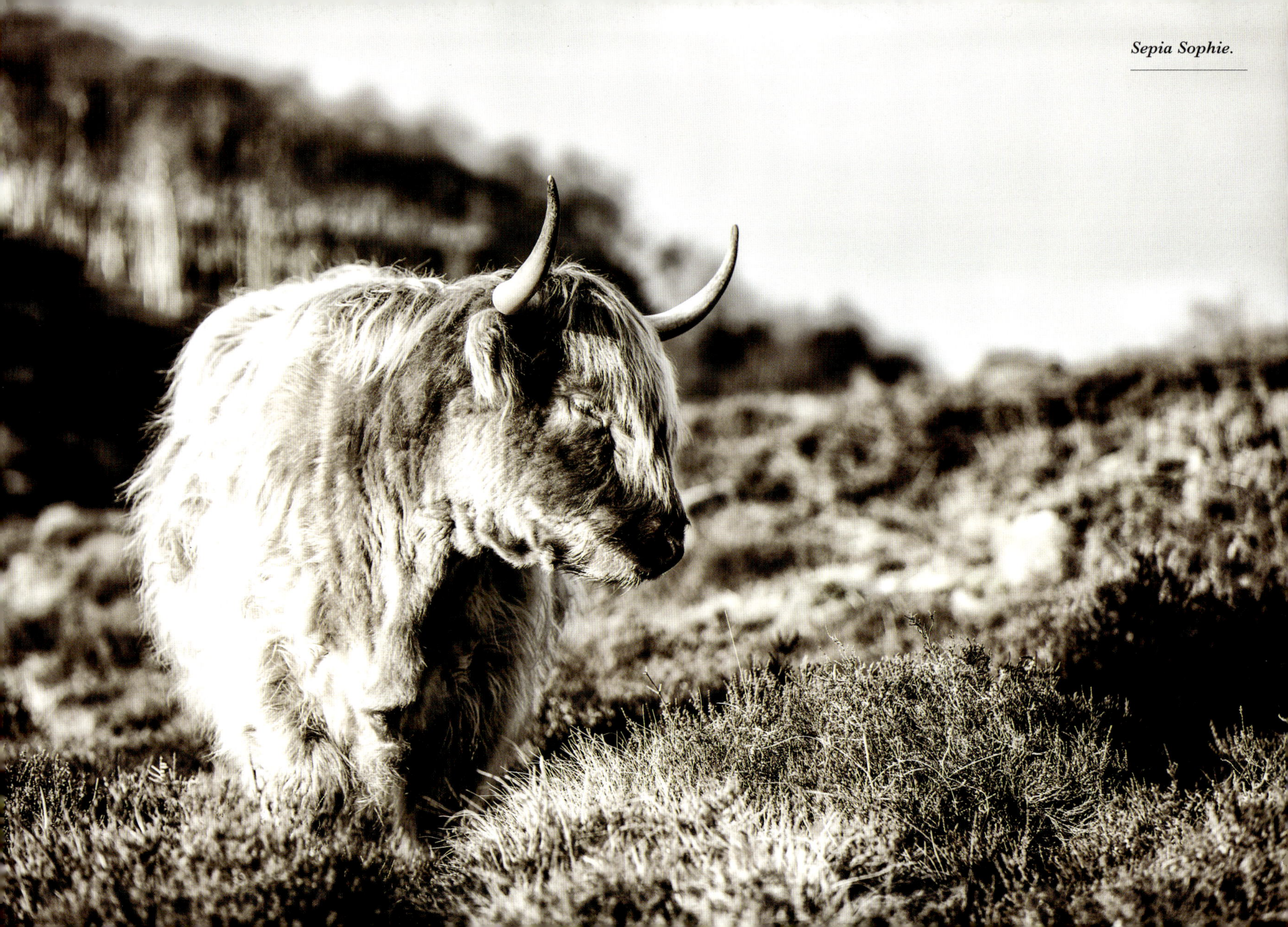
Sepia Sophie.

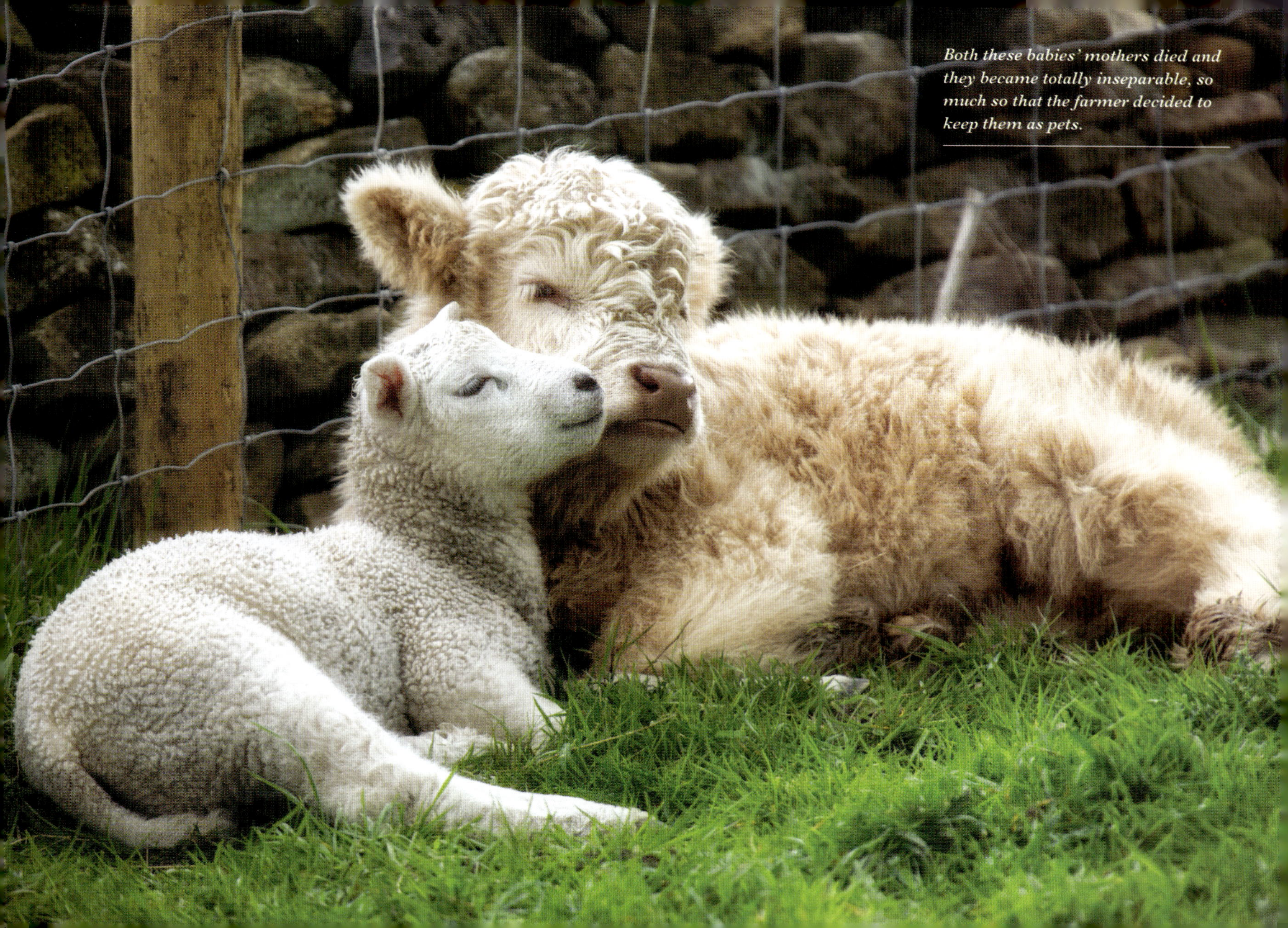

Both these babies' mothers died and they became totally inseparable, so much so that the farmer decided to keep them as pets.

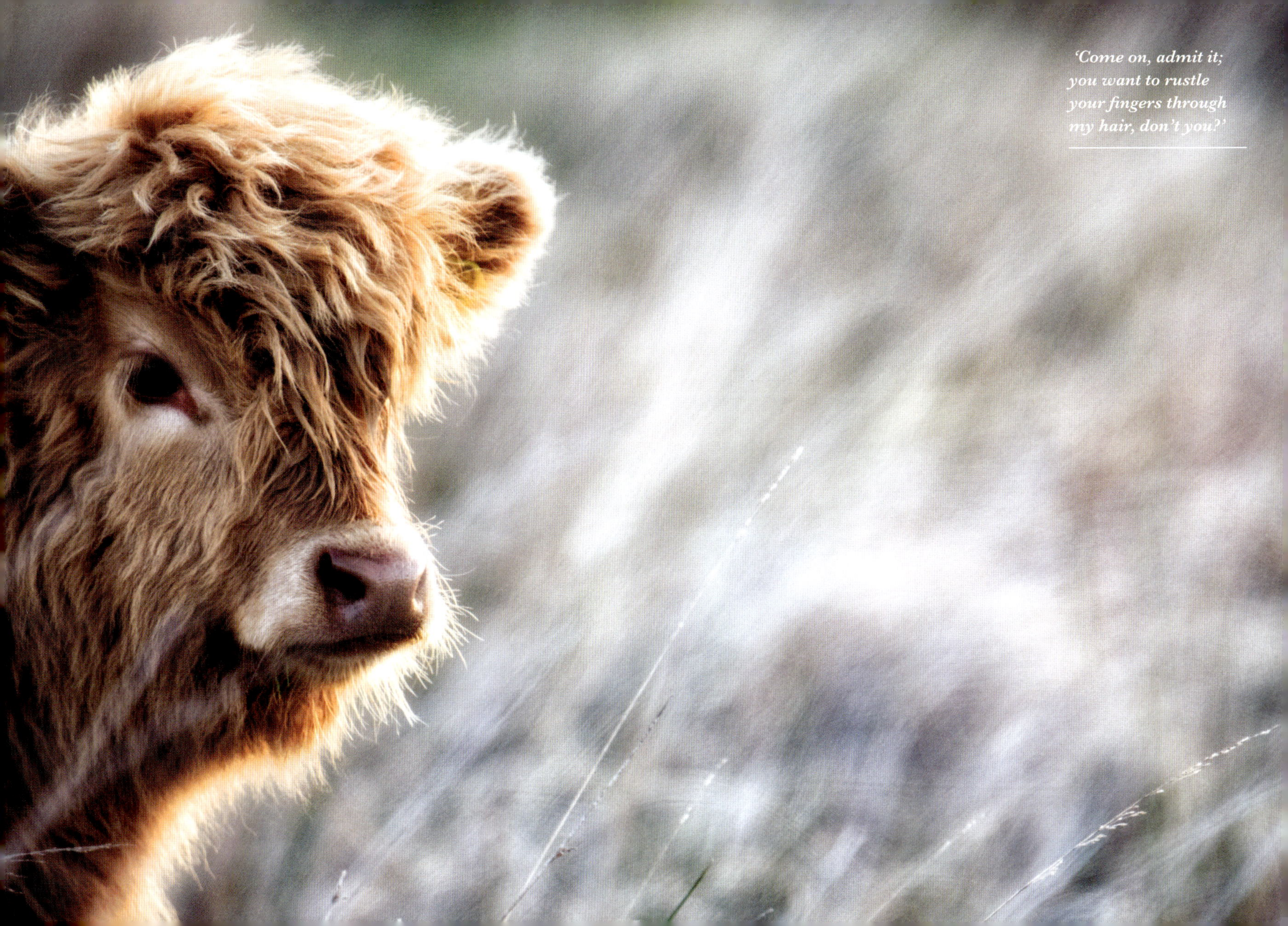
'Come on, admit it;
you want to rustle
your fingers through
my hair, don't you?'

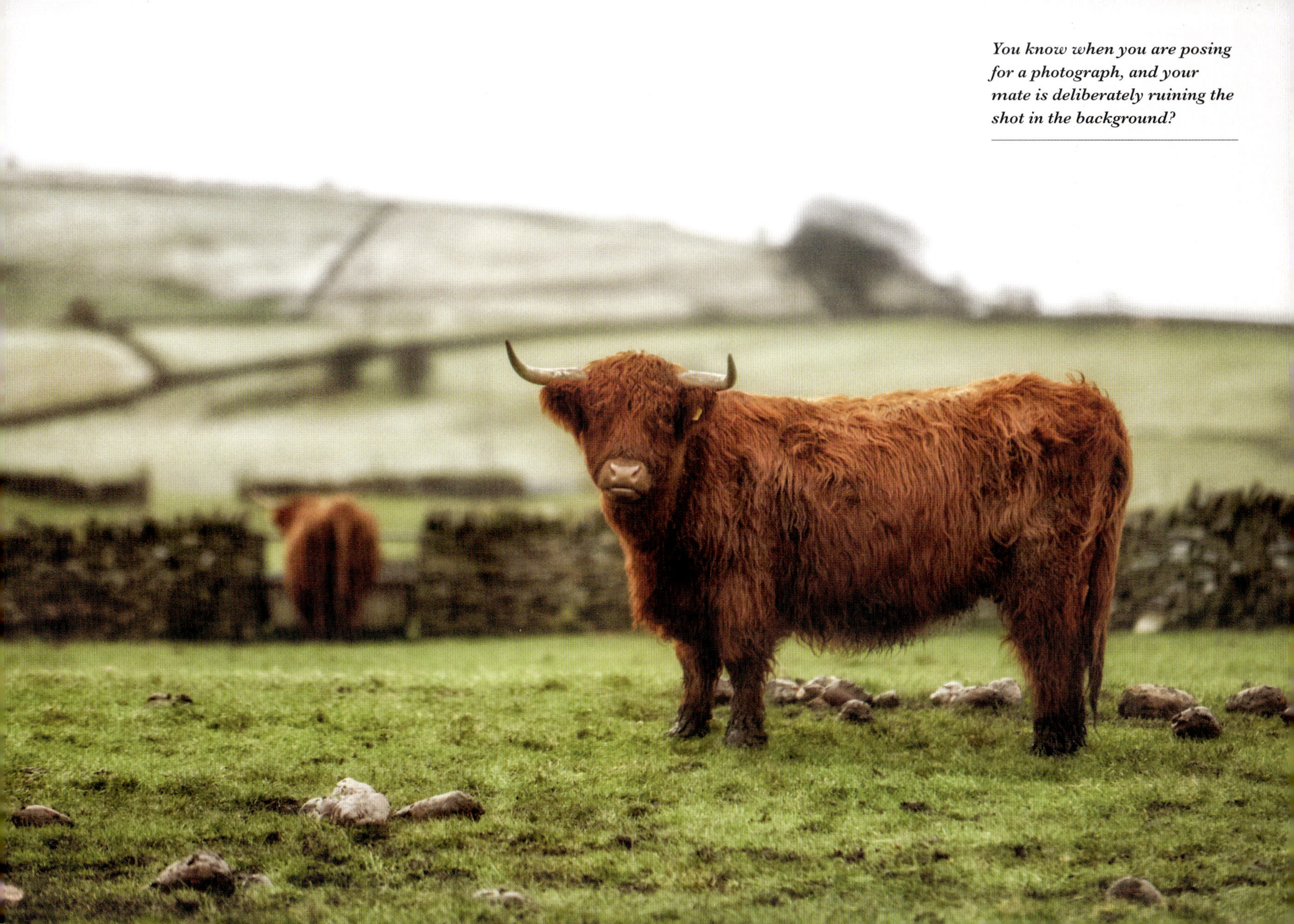

You know when you are posing
for a photograph, and your
mate is deliberately ruining the
shot in the background?

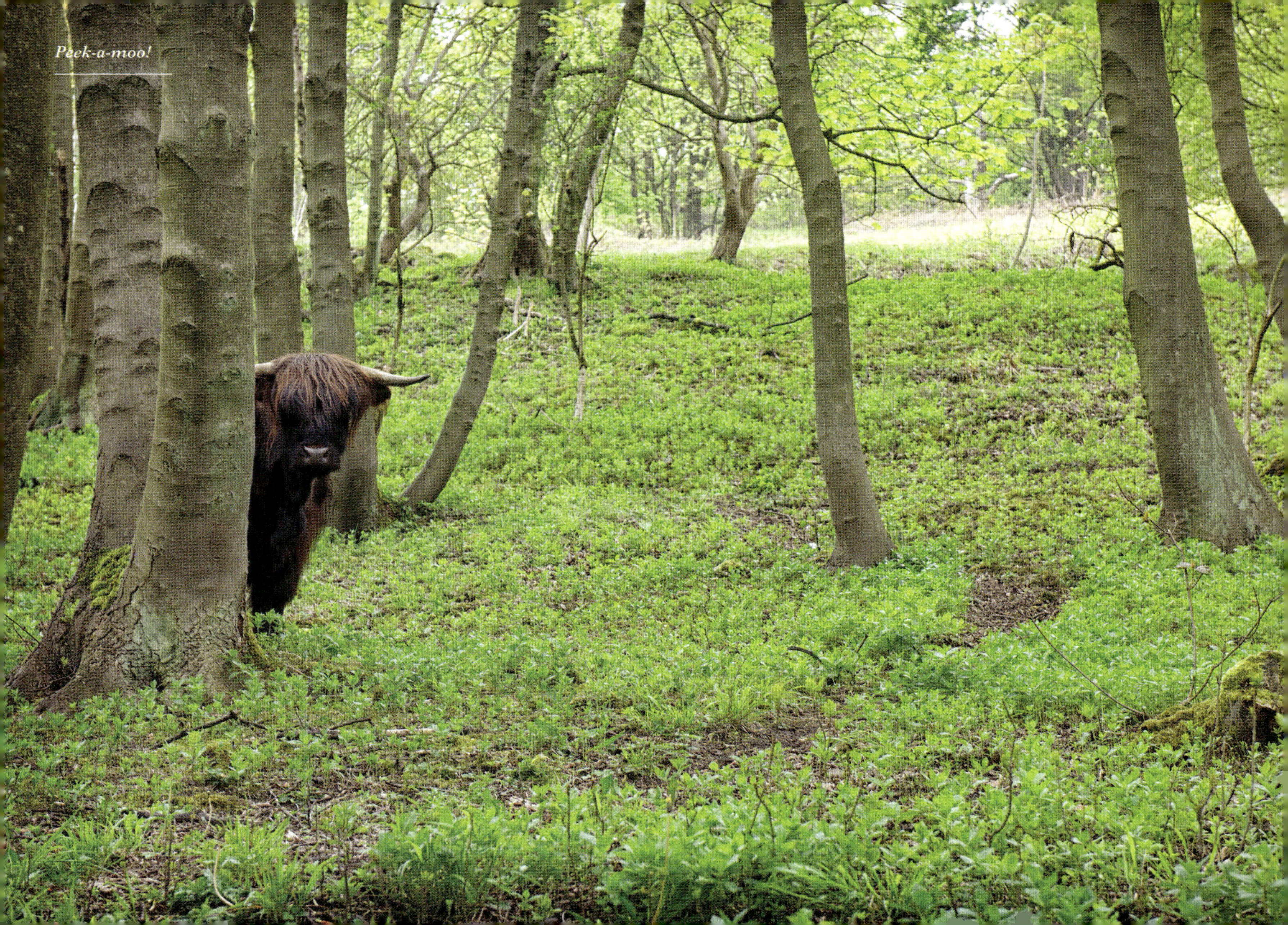

Peek-a-moo!

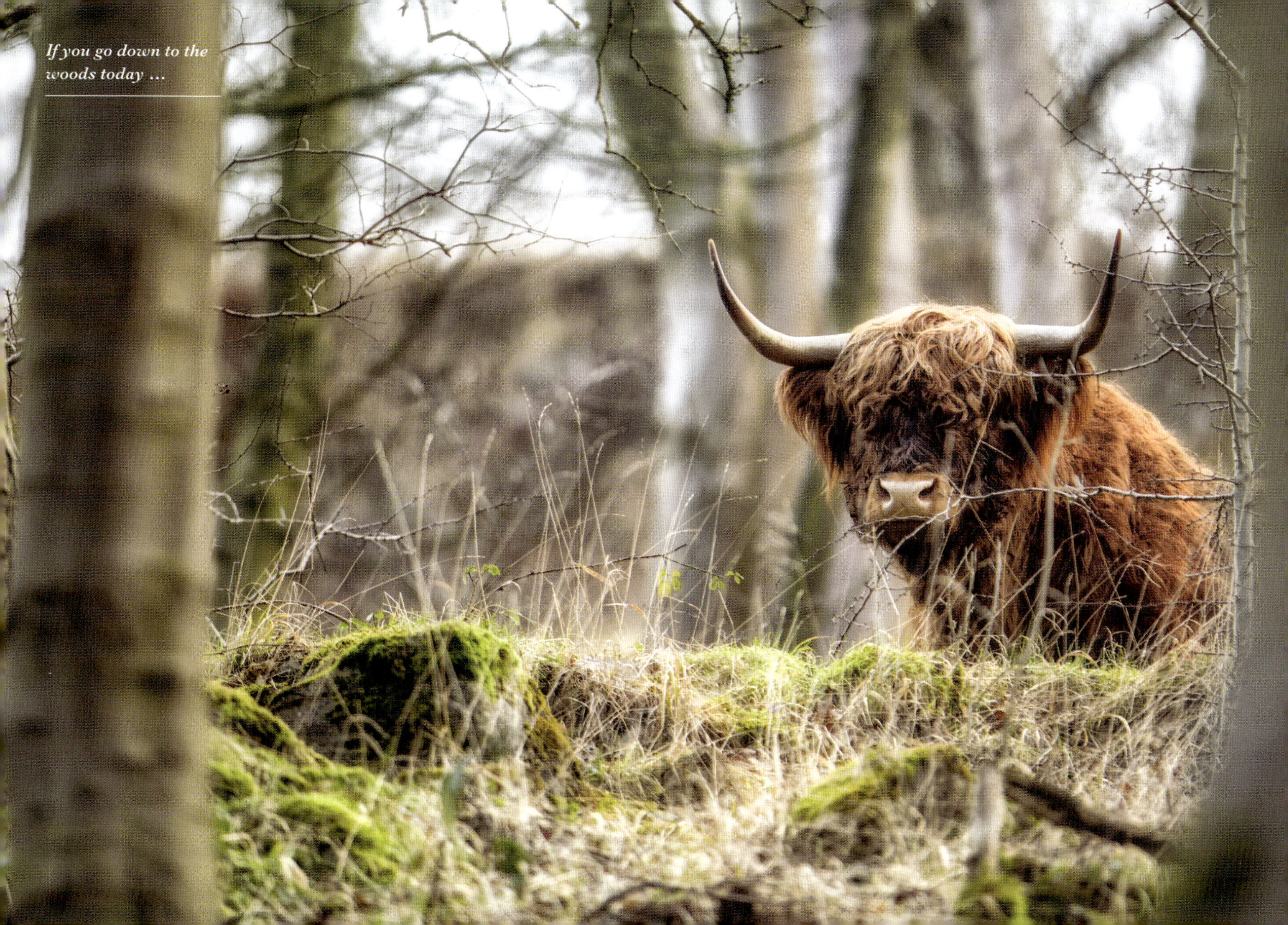
If you go down to the
woods today …

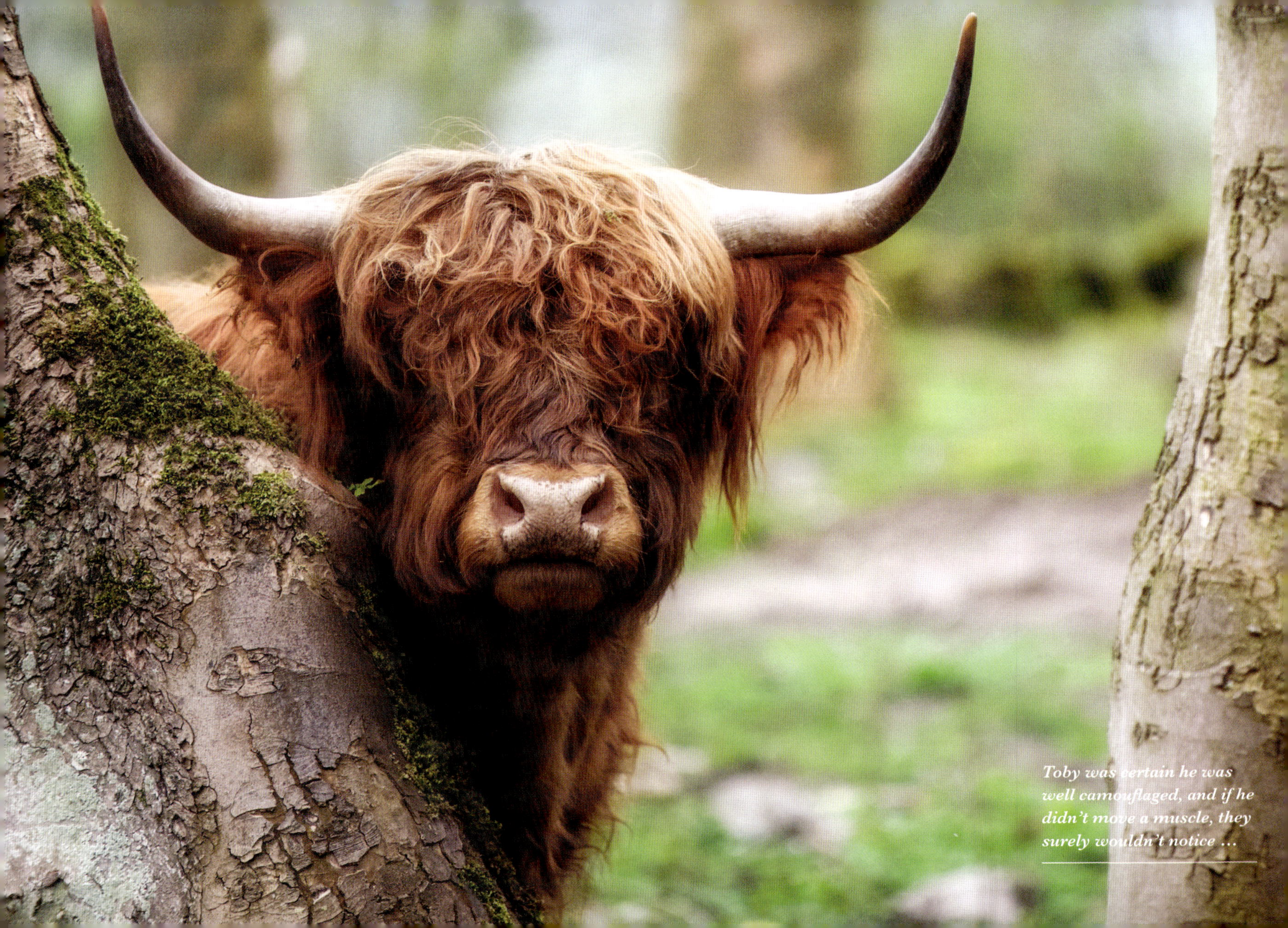

Toby was certain he was
well camouflaged, and if he
didn't move a muscle, they
surely wouldn't notice …

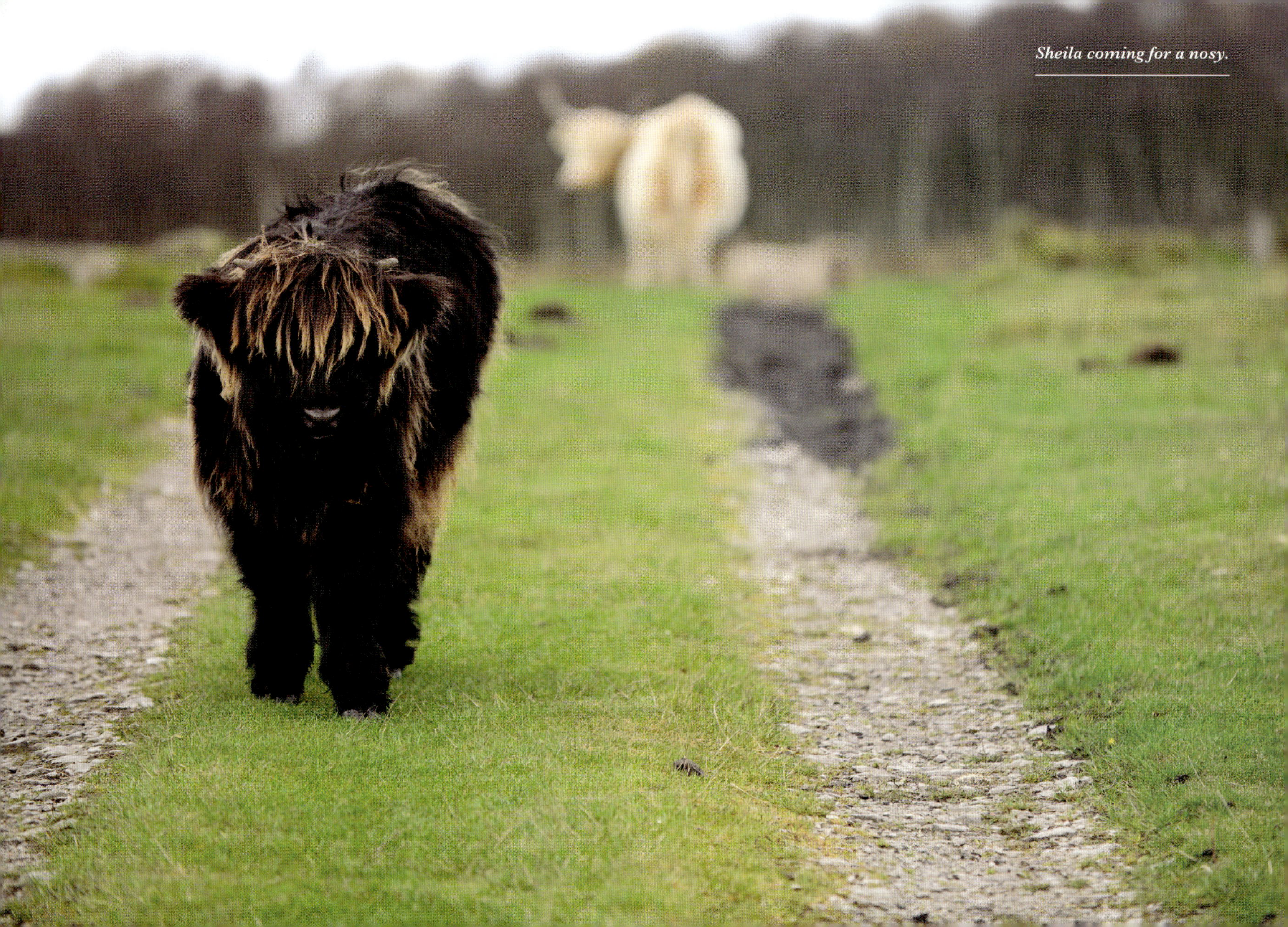
Sheila coming for a nosy.

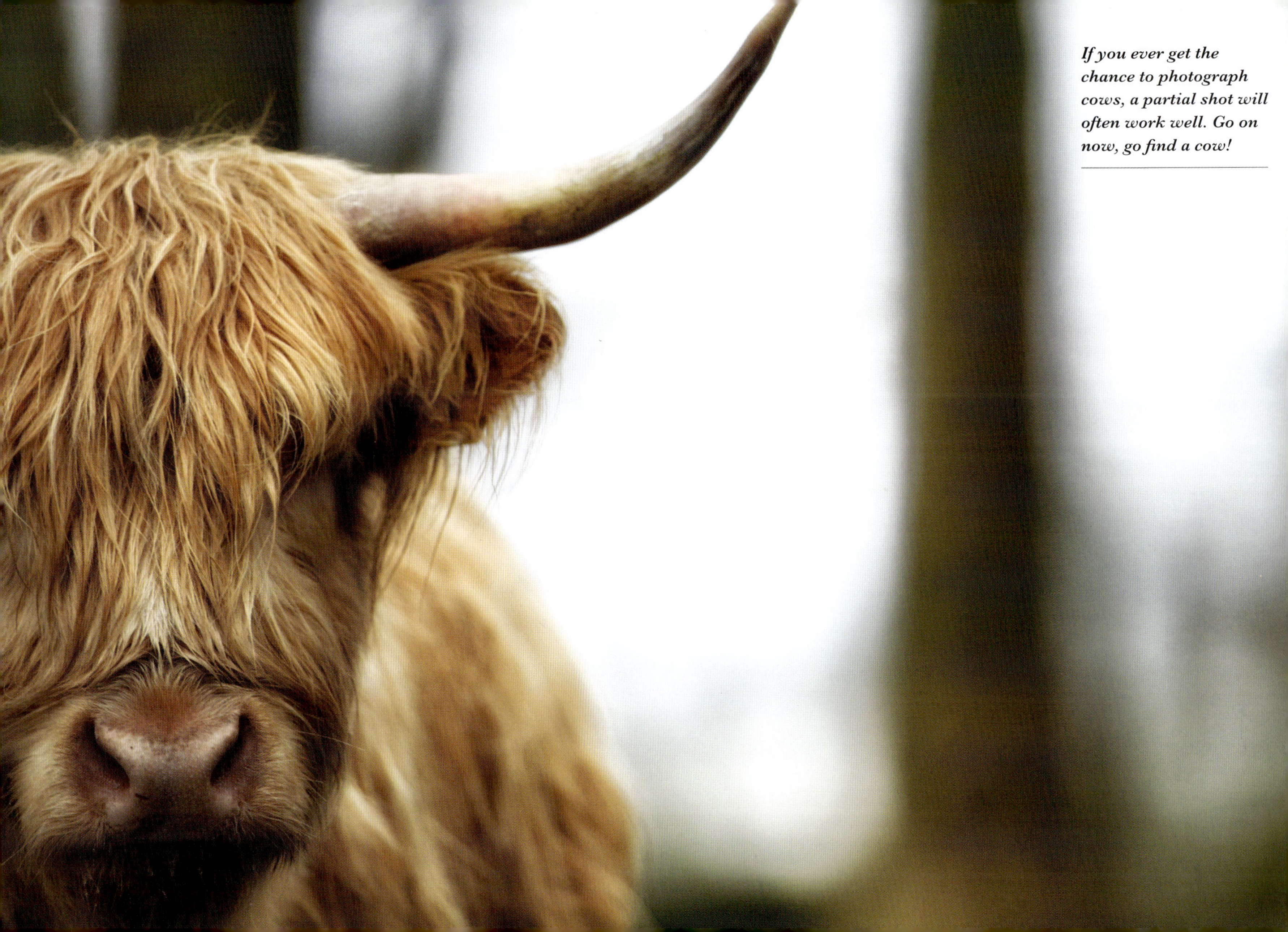
If you ever get the
chance to photograph
cows, a partial shot will
often work well. Go on
now, go find a cow!

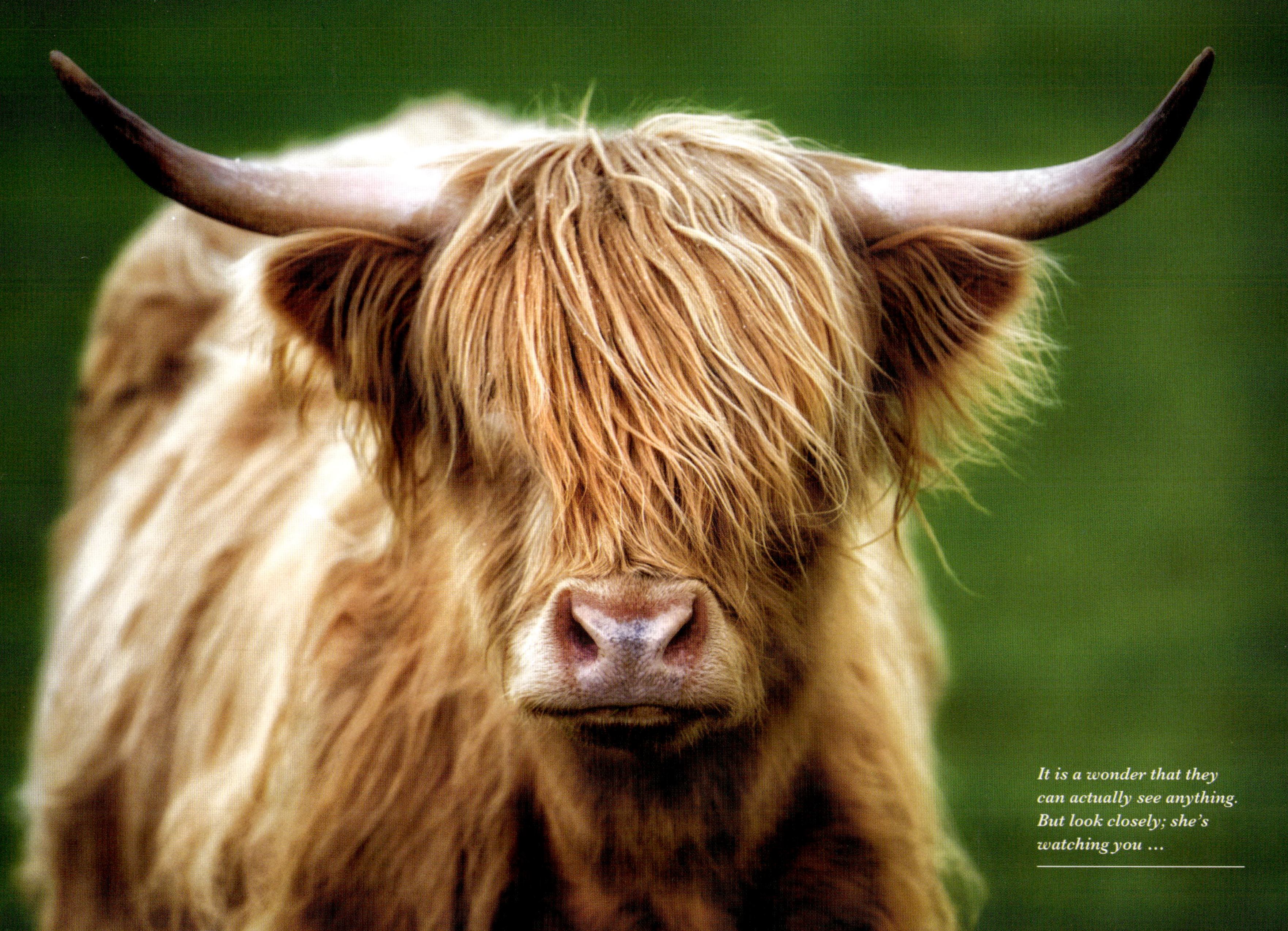

It is a wonder that they
can actually see anything.
But look closely; she's
watching you …

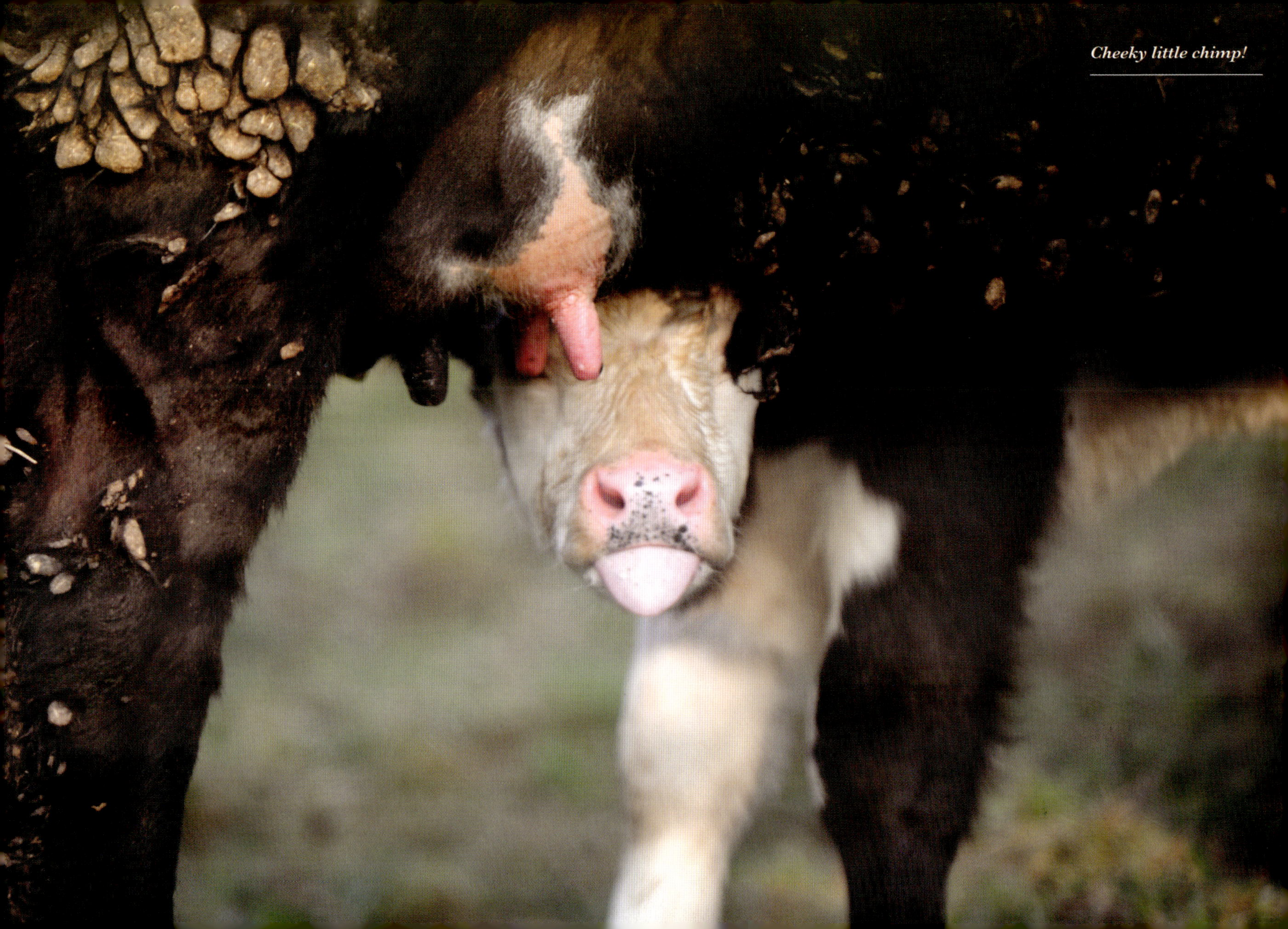

Cheeky little chimp!

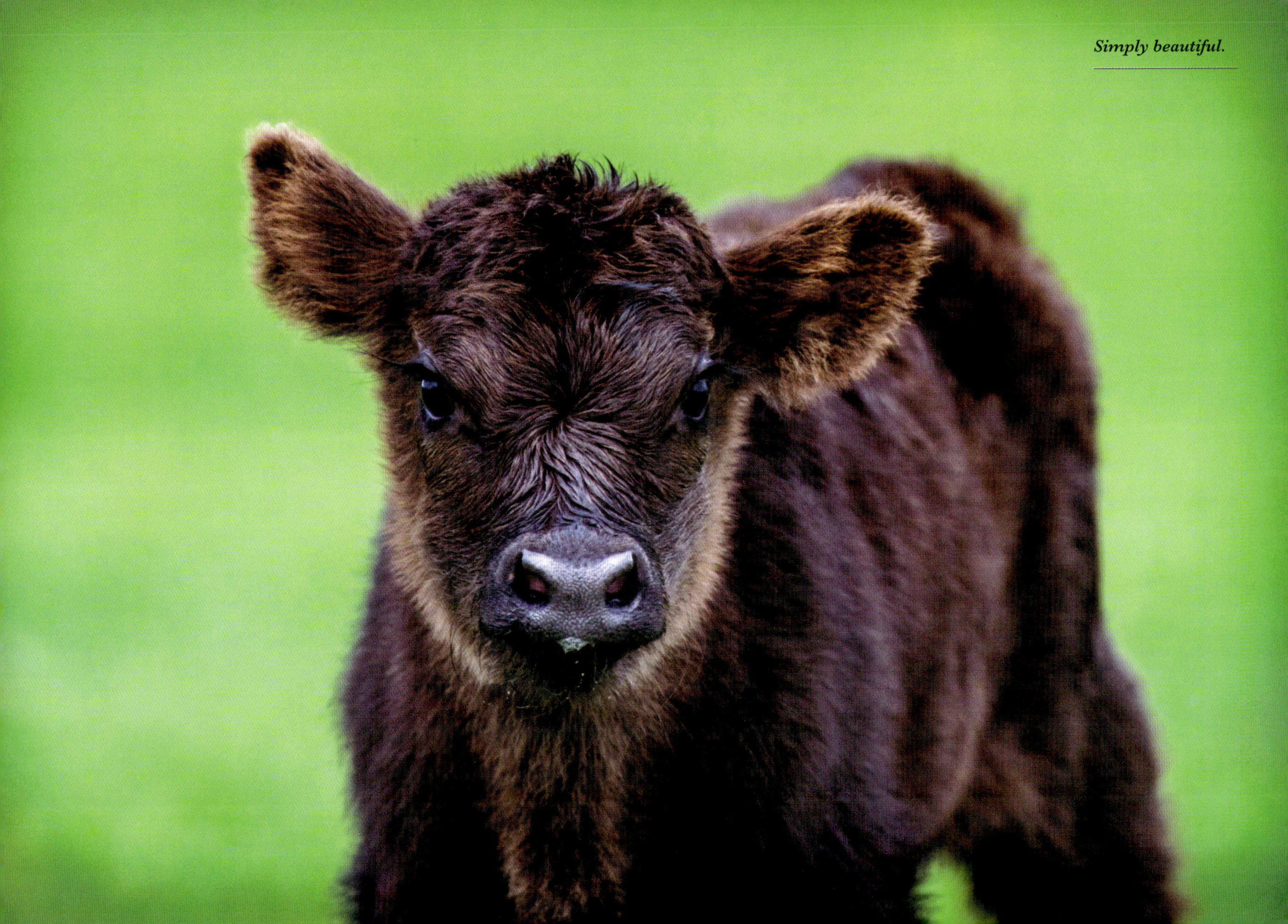
Simply beautiful.

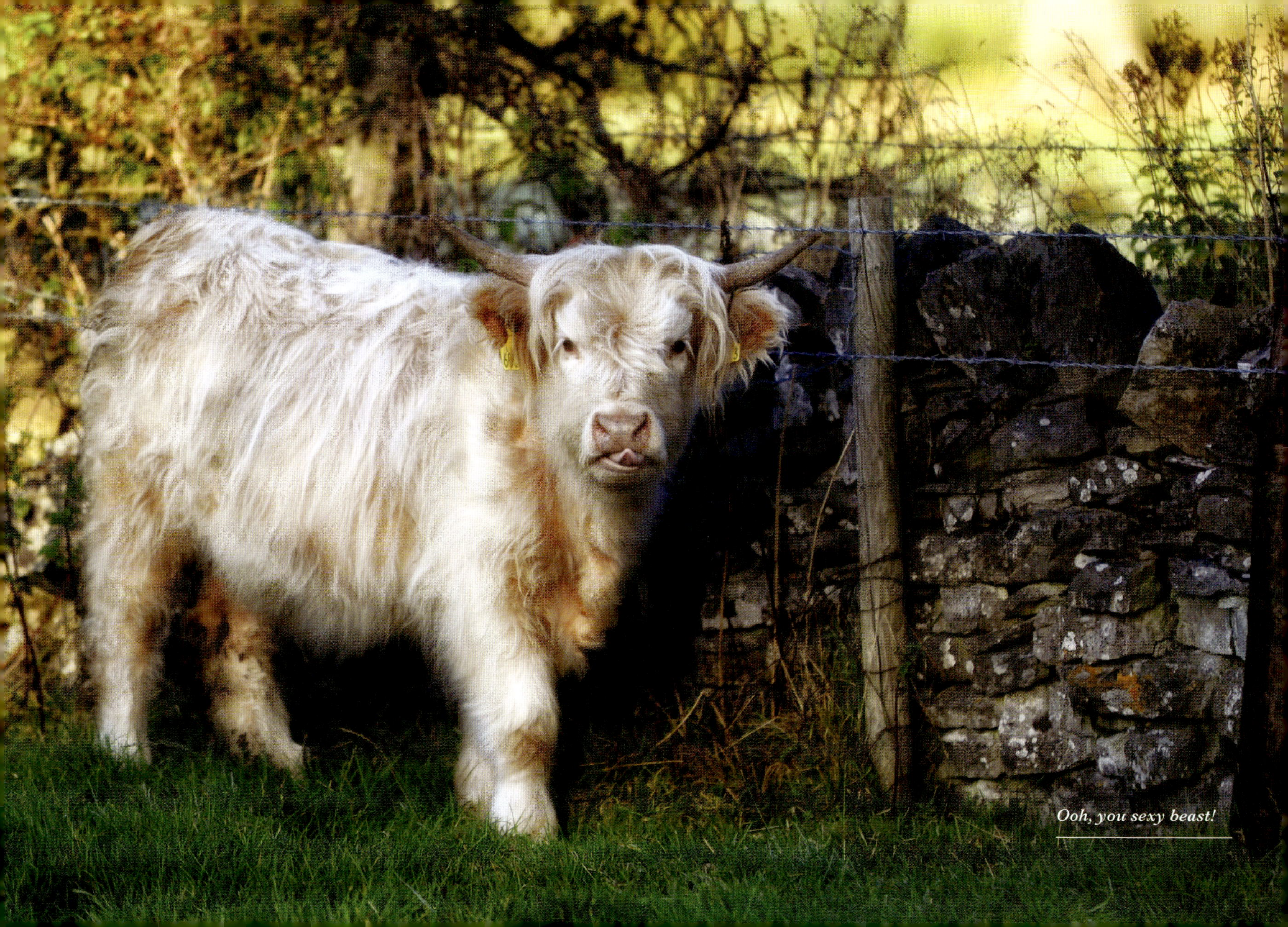

Ooh, you sexy beast!

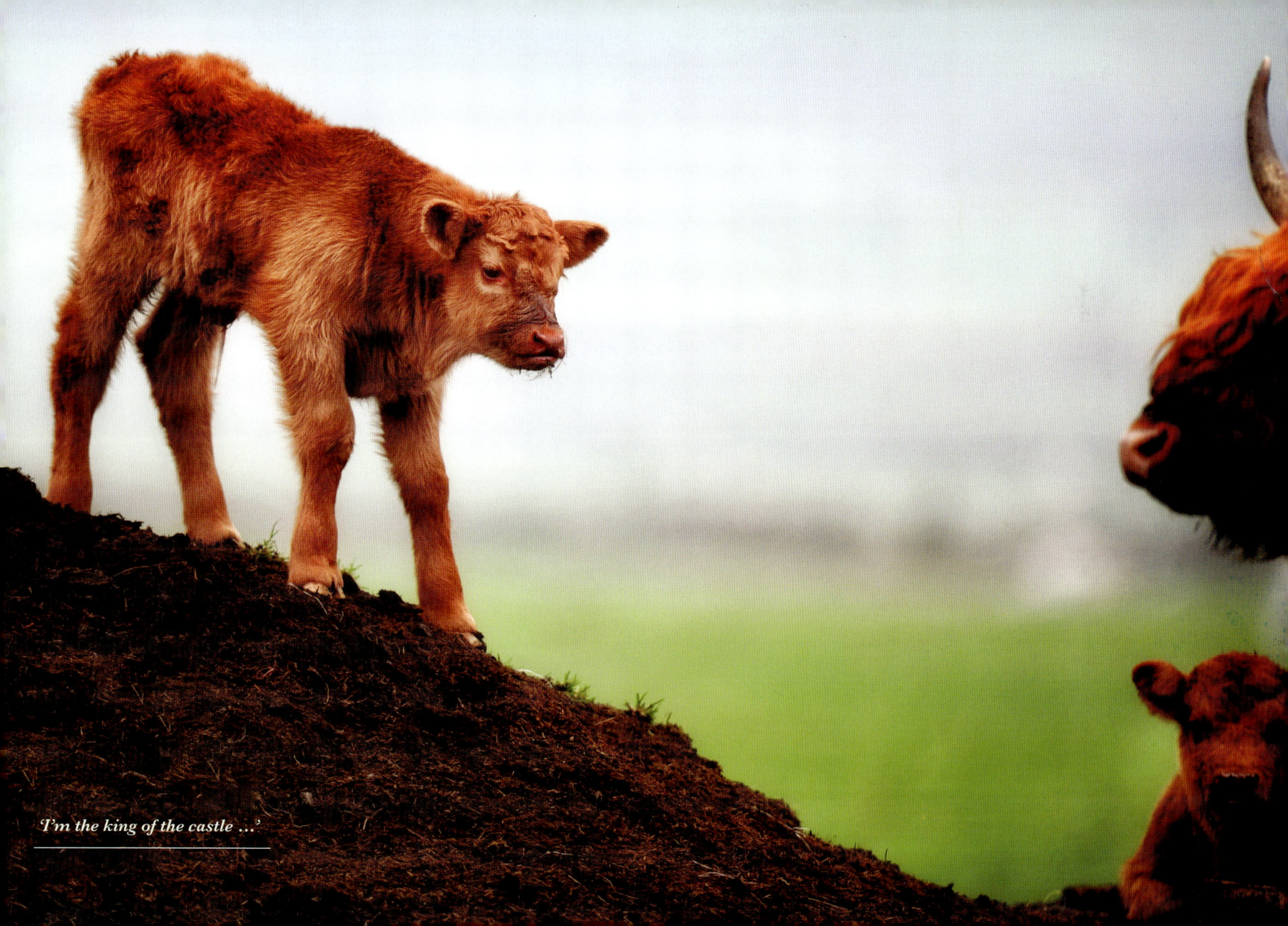

'I'm the king of the castle …'

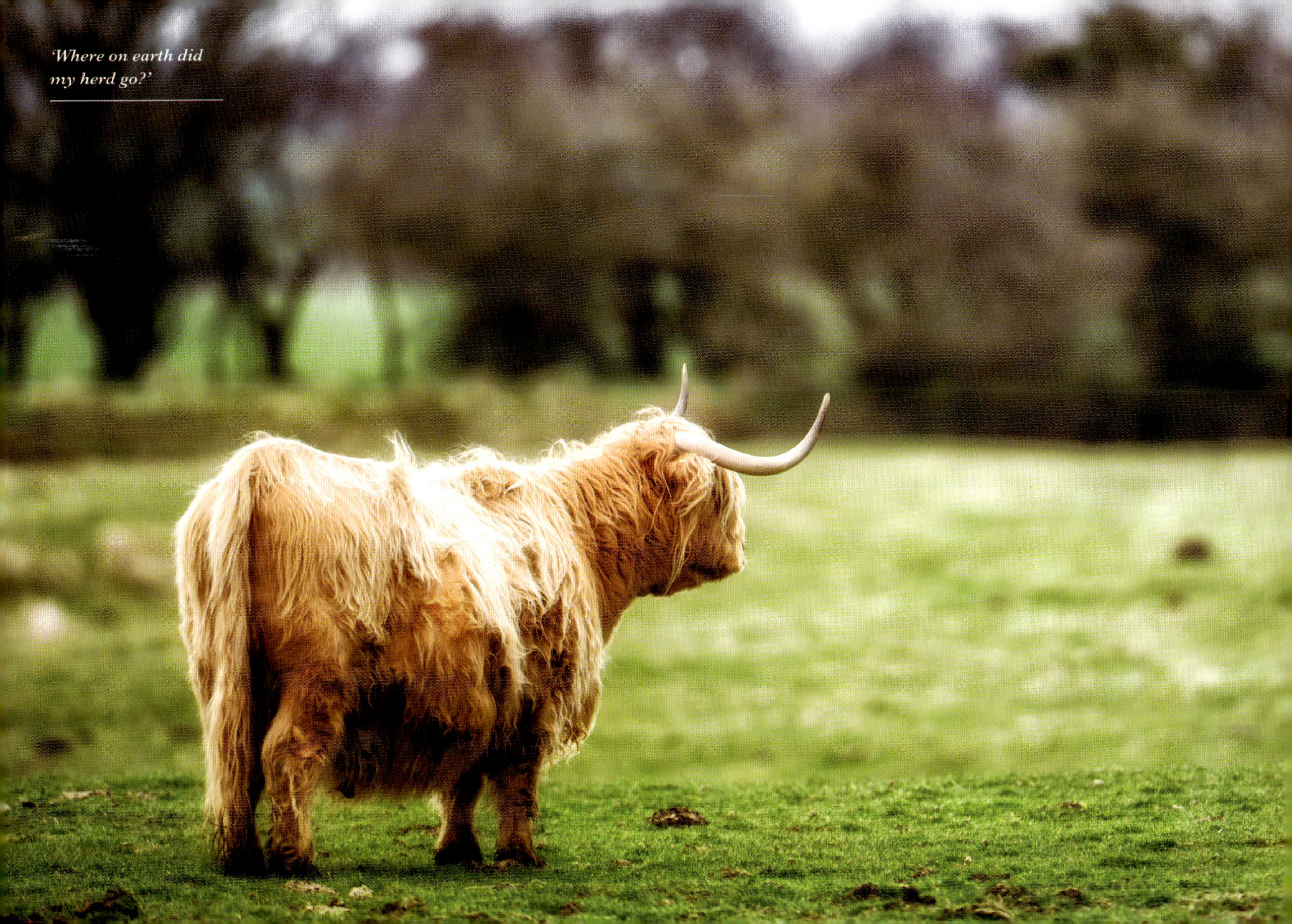

'Where on earth did my herd go?'

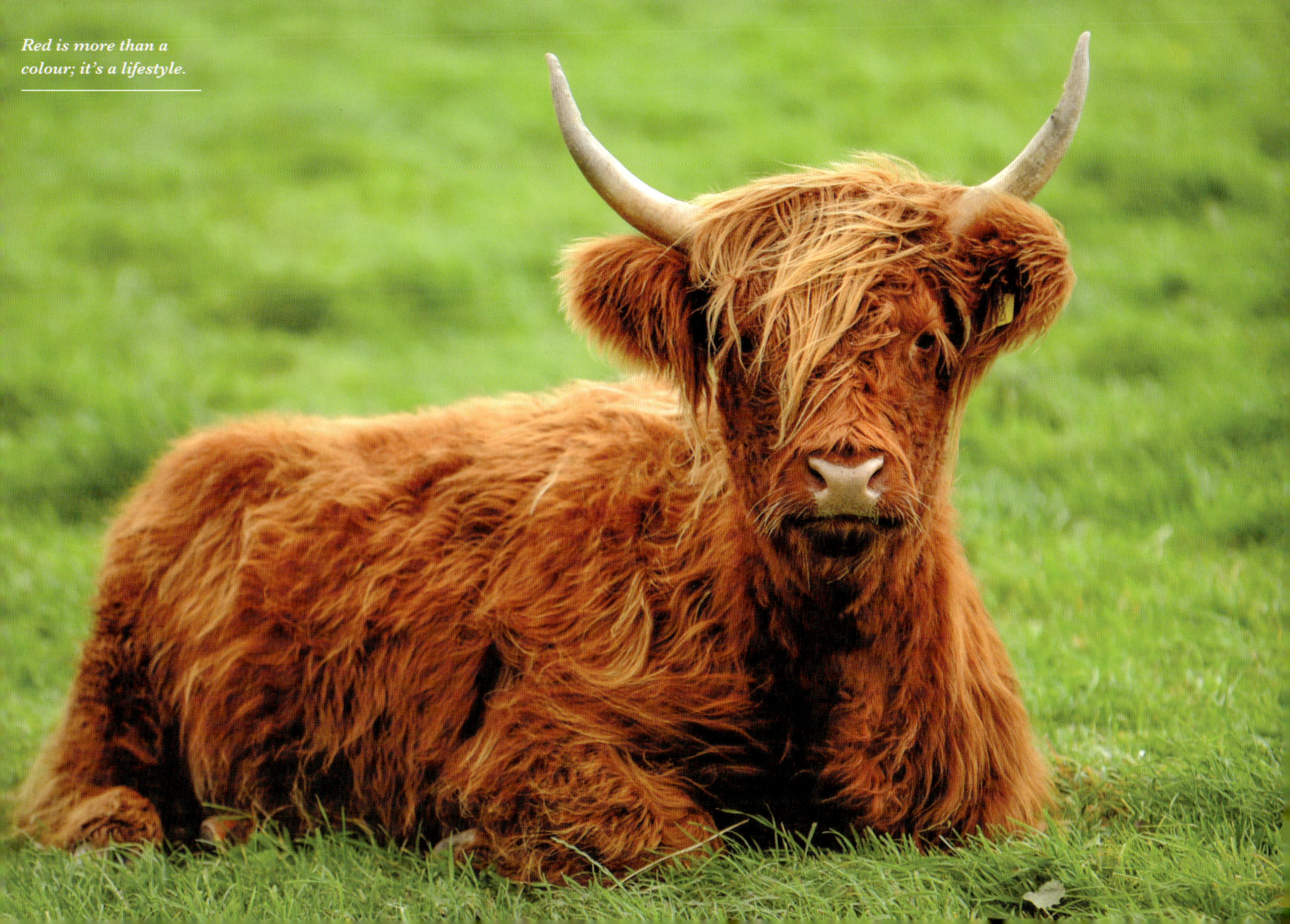
Red is more than a
colour; it's a lifestyle.

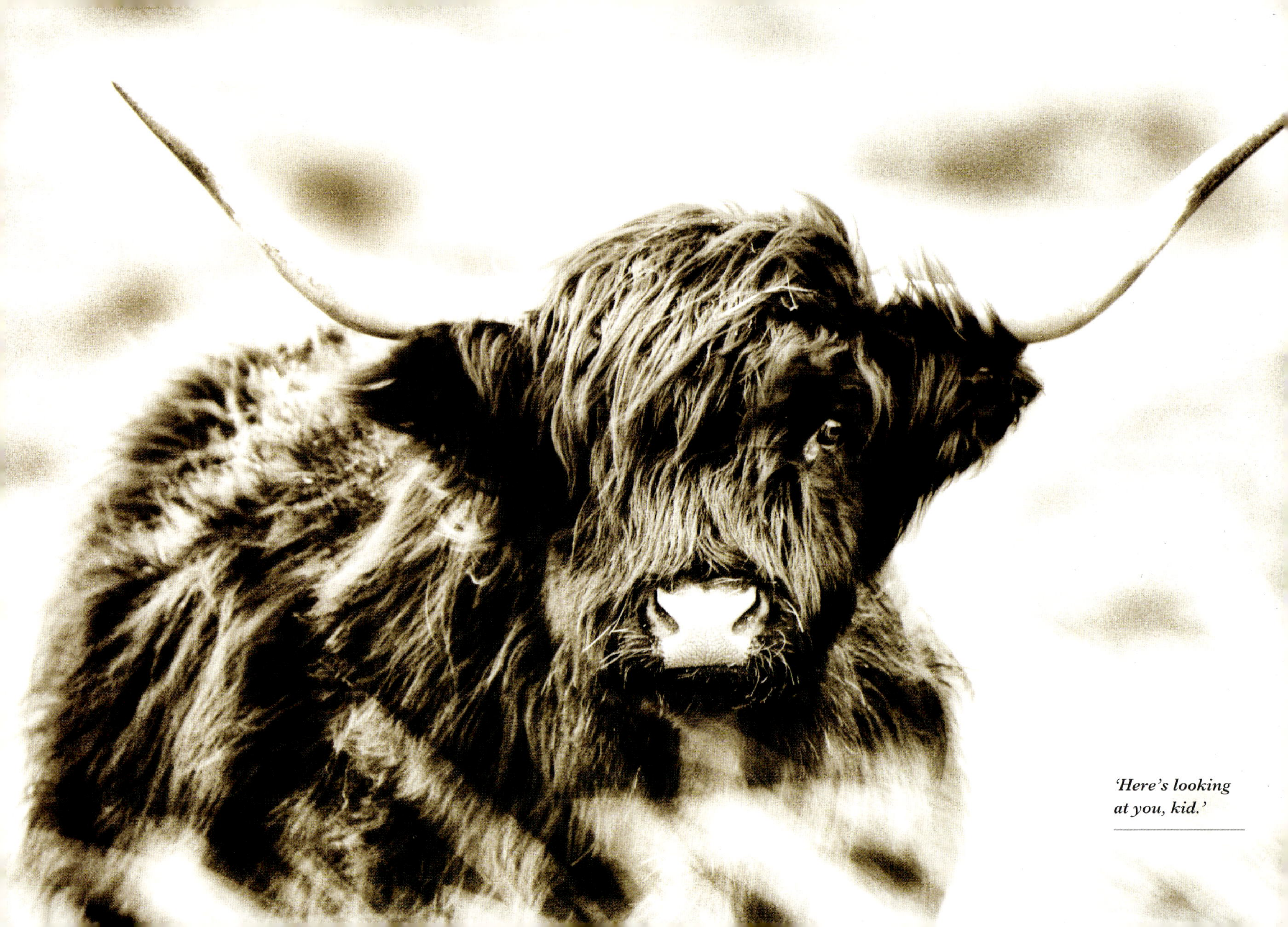

'Here's looking at you, kid.'

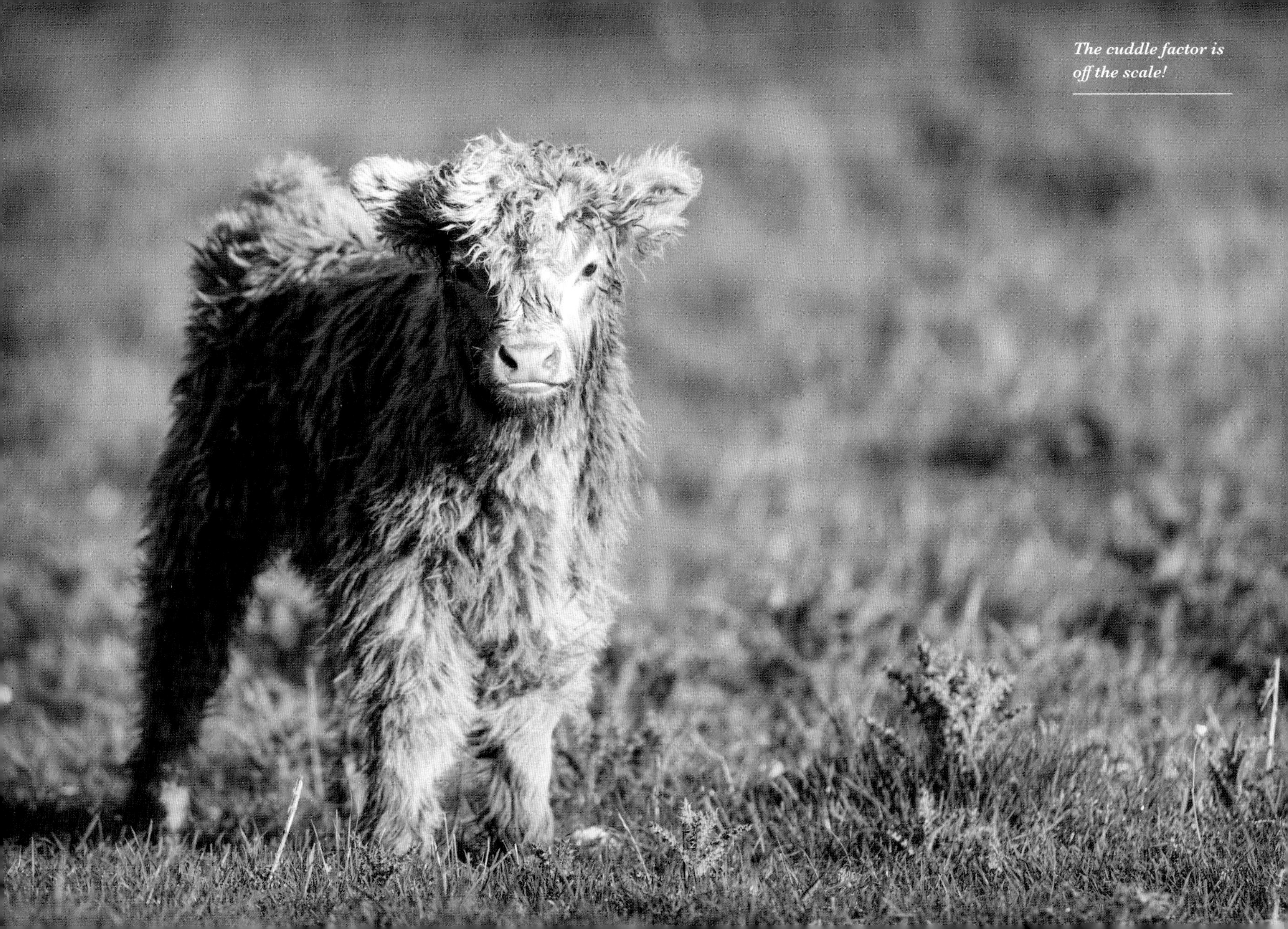
The cuddle factor is
off the scale!

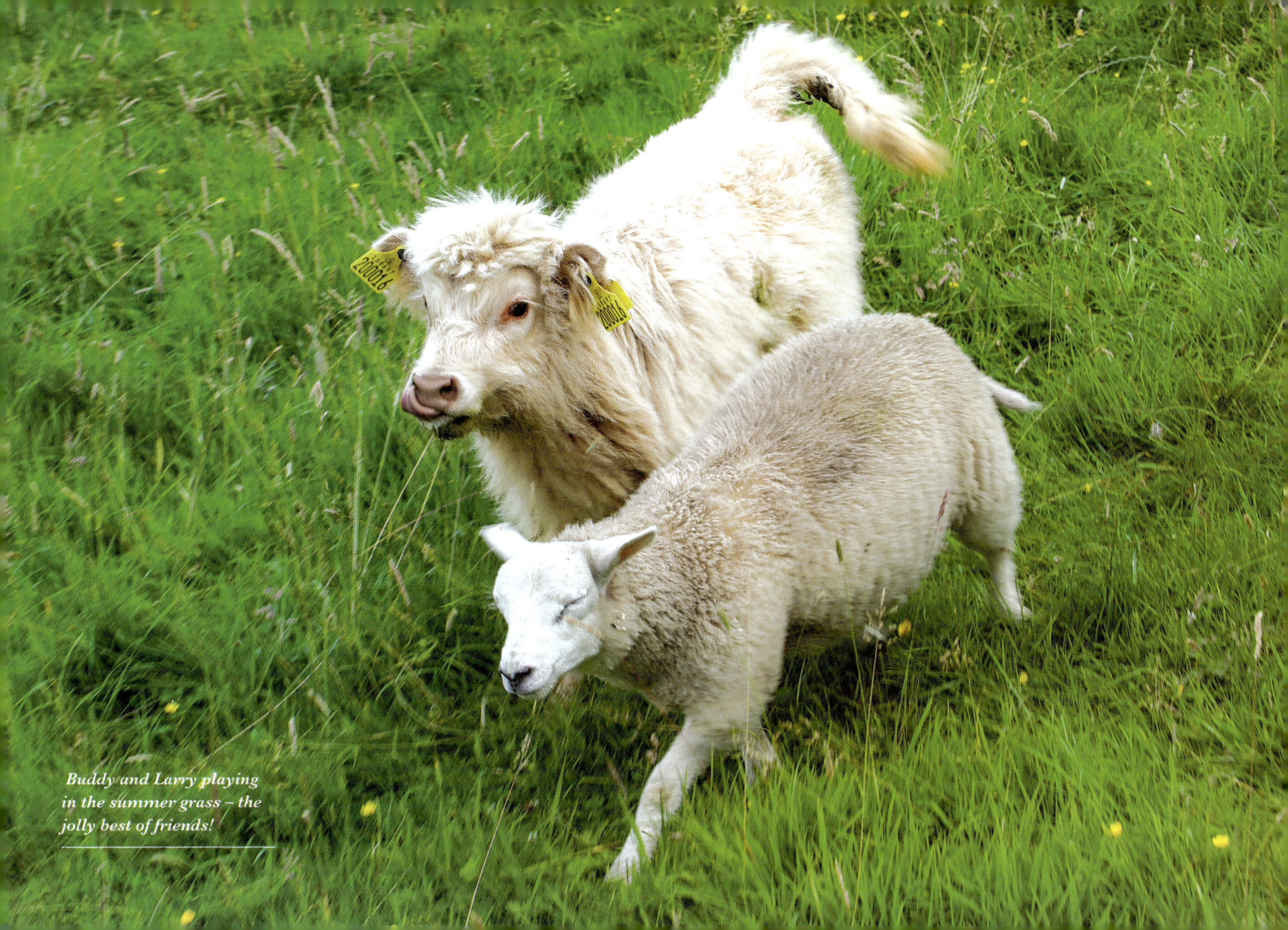

Buddy and Larry playing in the summer grass – the jolly best of friends!

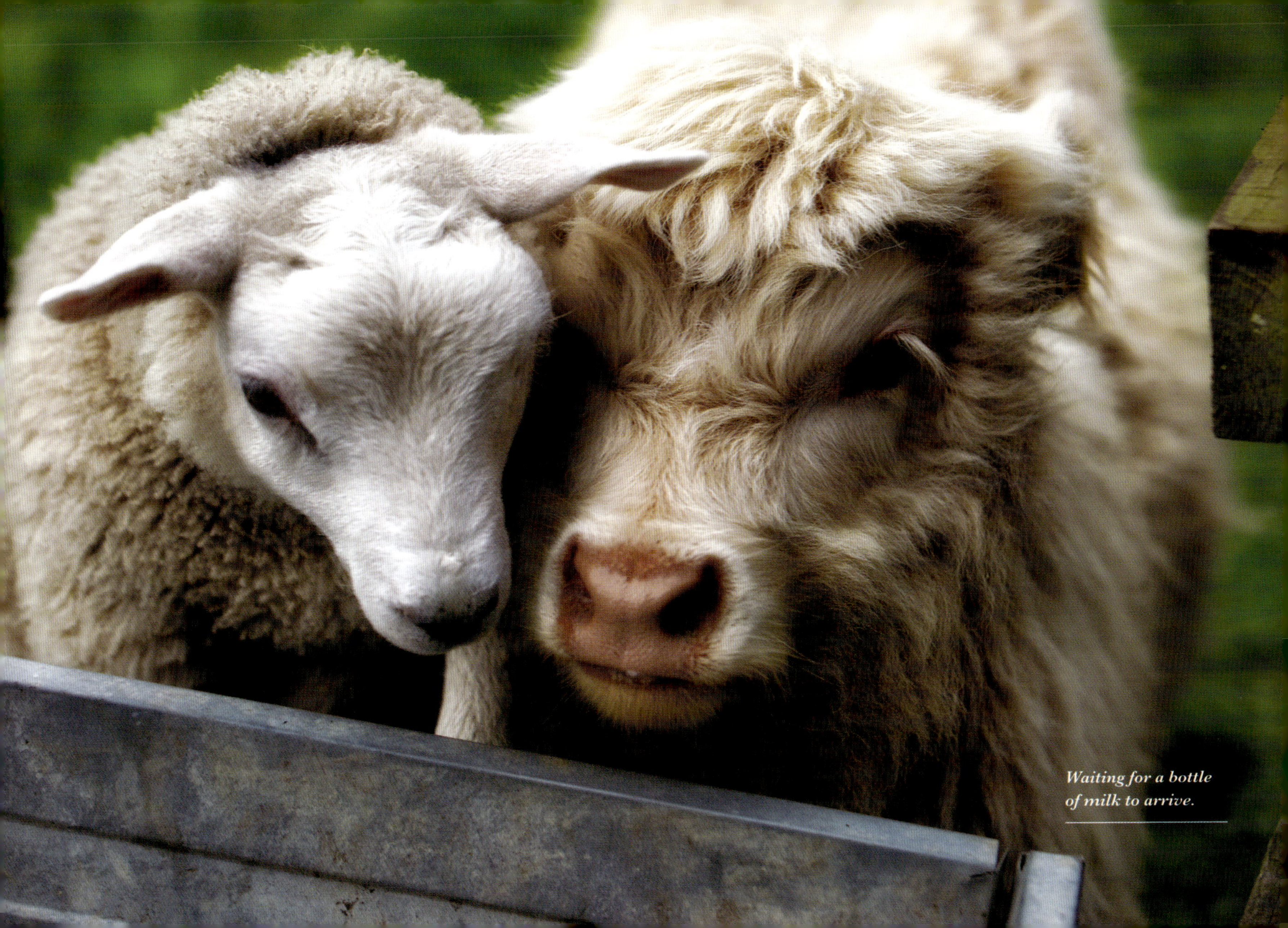

Waiting for a bottle
of milk to arrive.

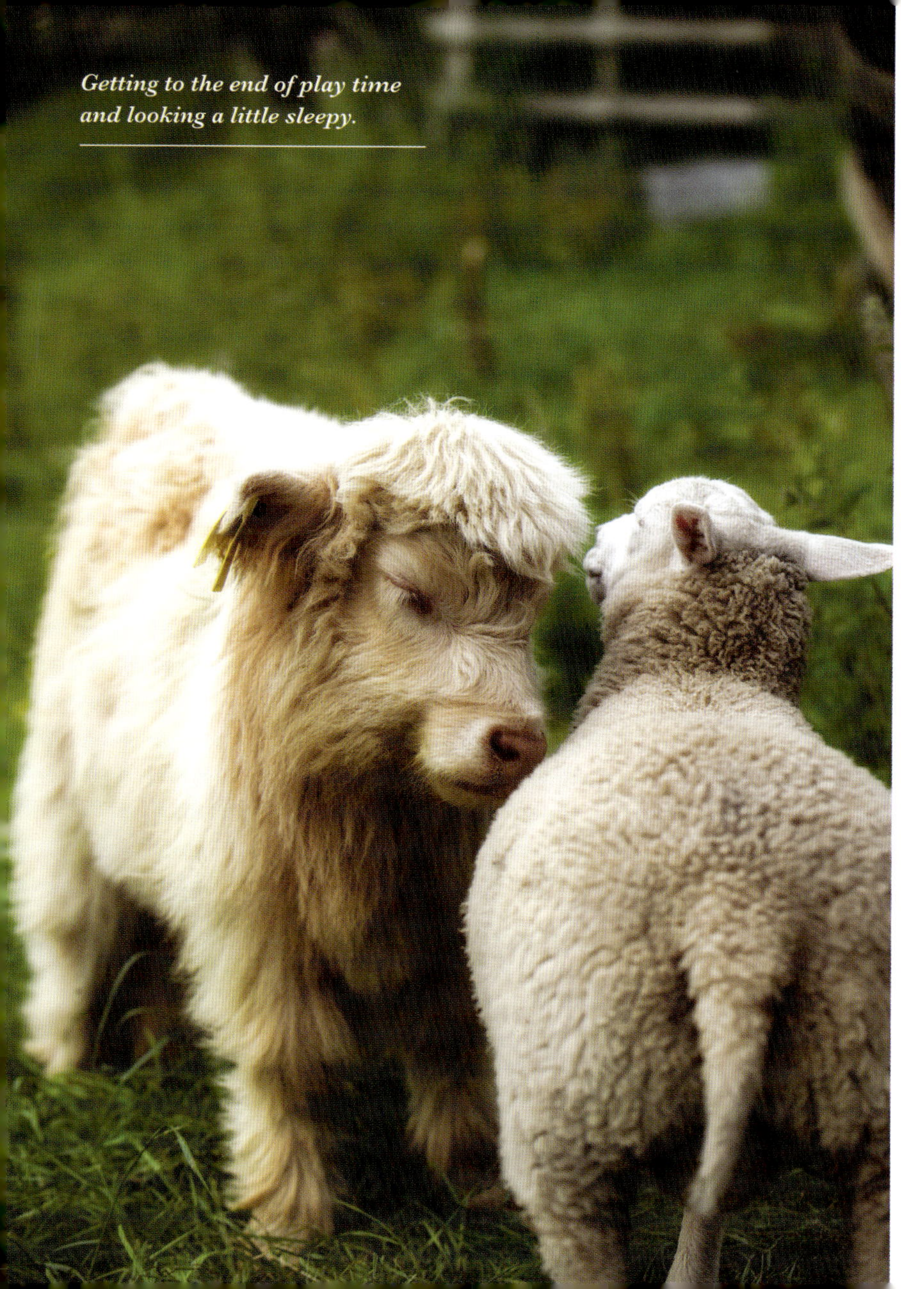

Getting to the end of play time
and looking a little sleepy.

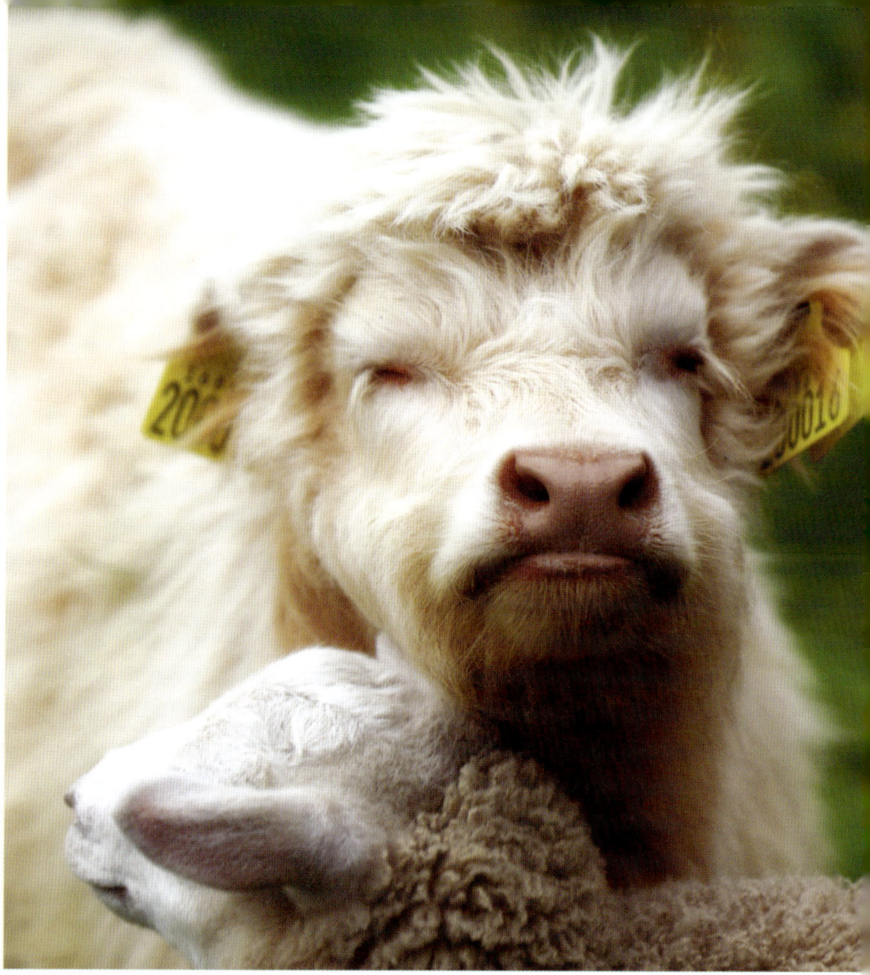

'Oh yes, that's the spot … just behind the ear … that feels so good …'

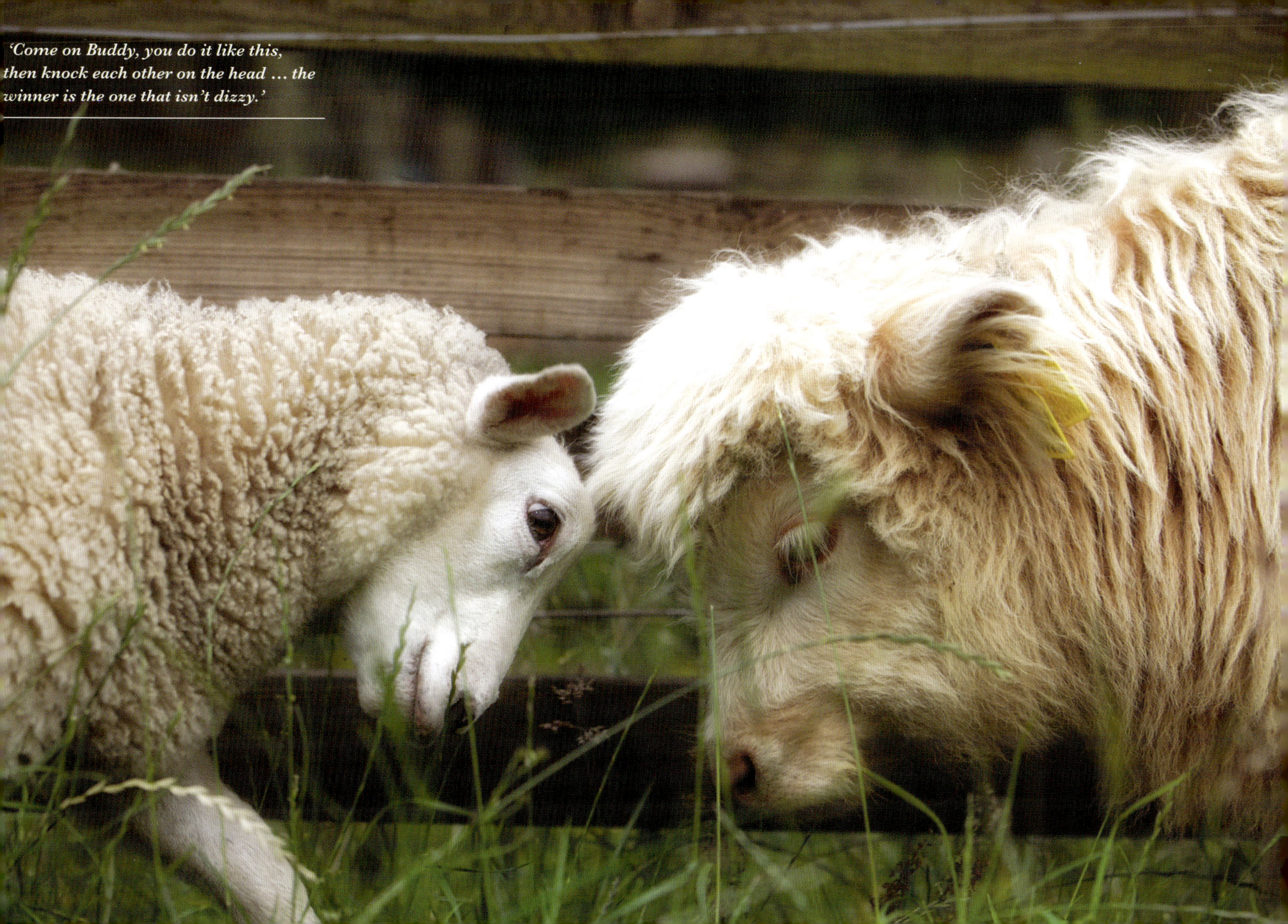
'Come on Buddy, you do it like this,
then knock each other on the head … the
winner is the one that isn't dizzy.'

Fiona takes a
long walk for a
little calf …

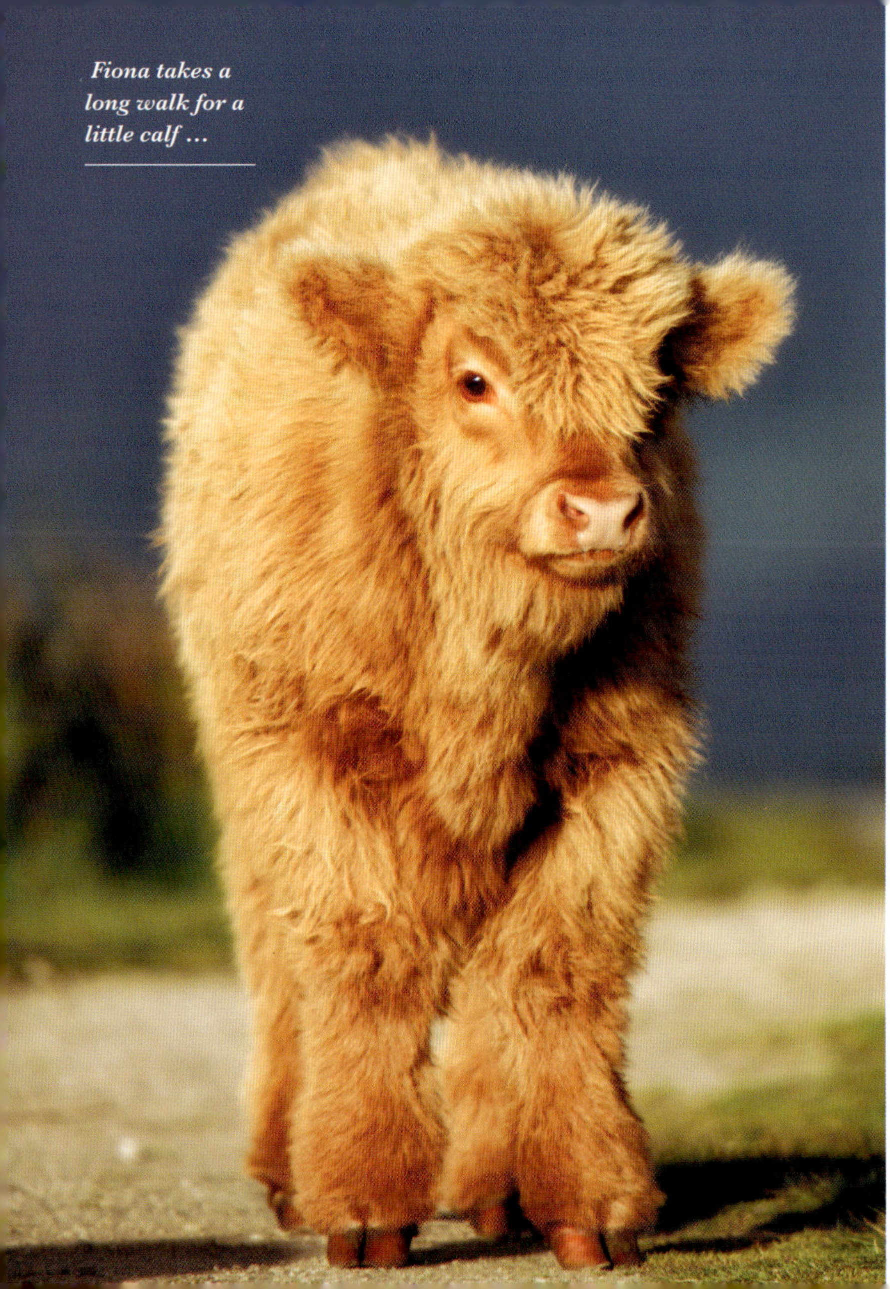

… and the hooves
are starting to hurt.

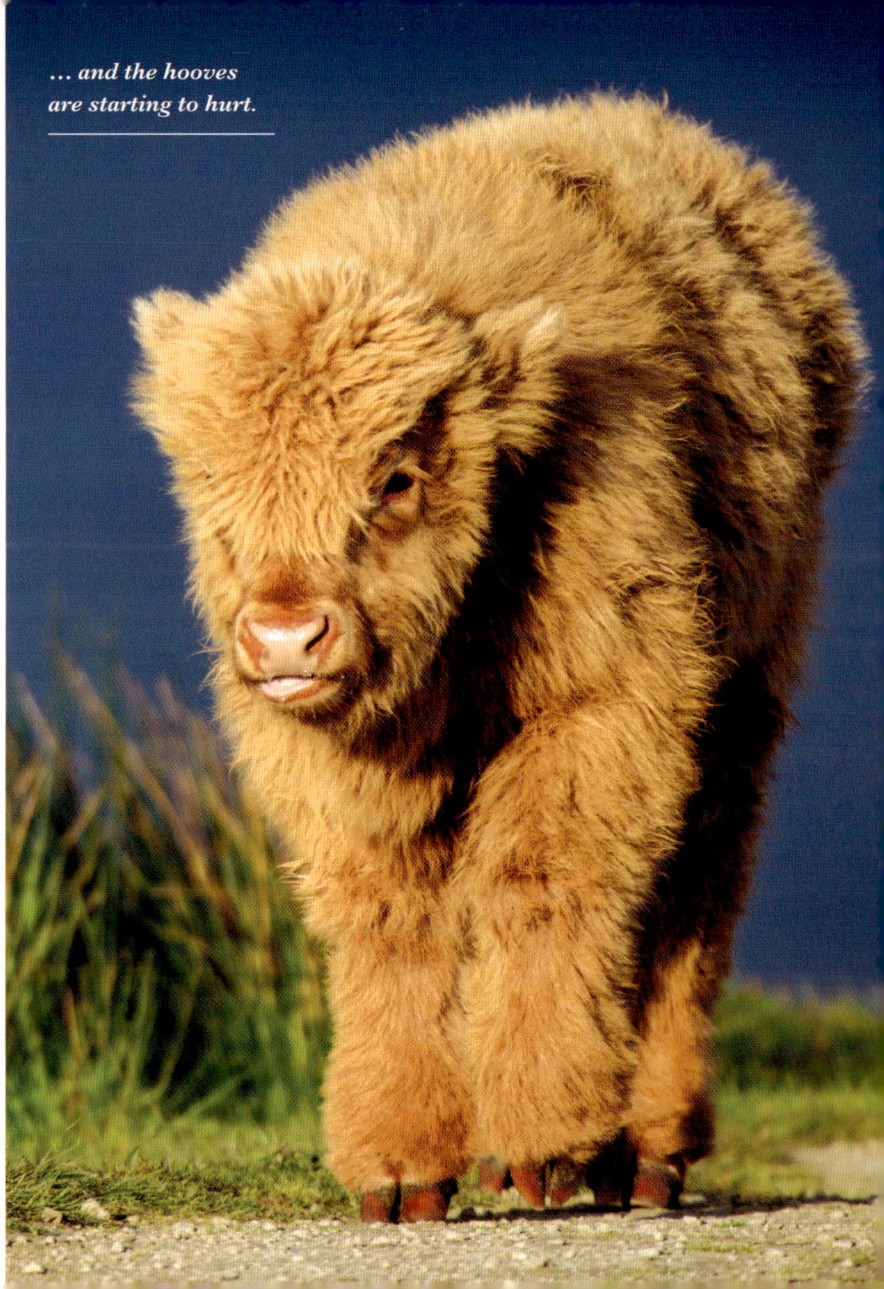

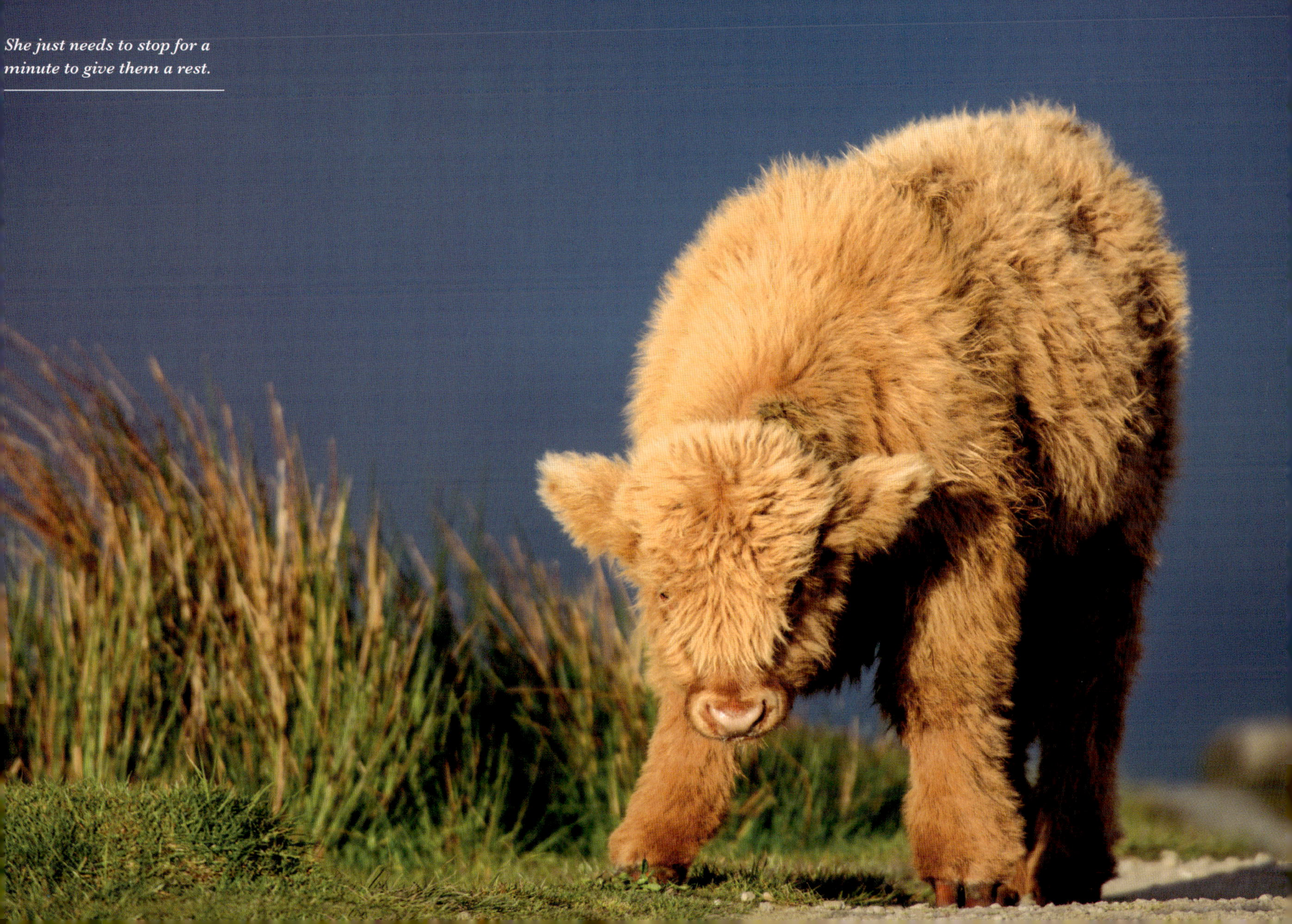
She just needs to stop for a
minute to give them a rest.

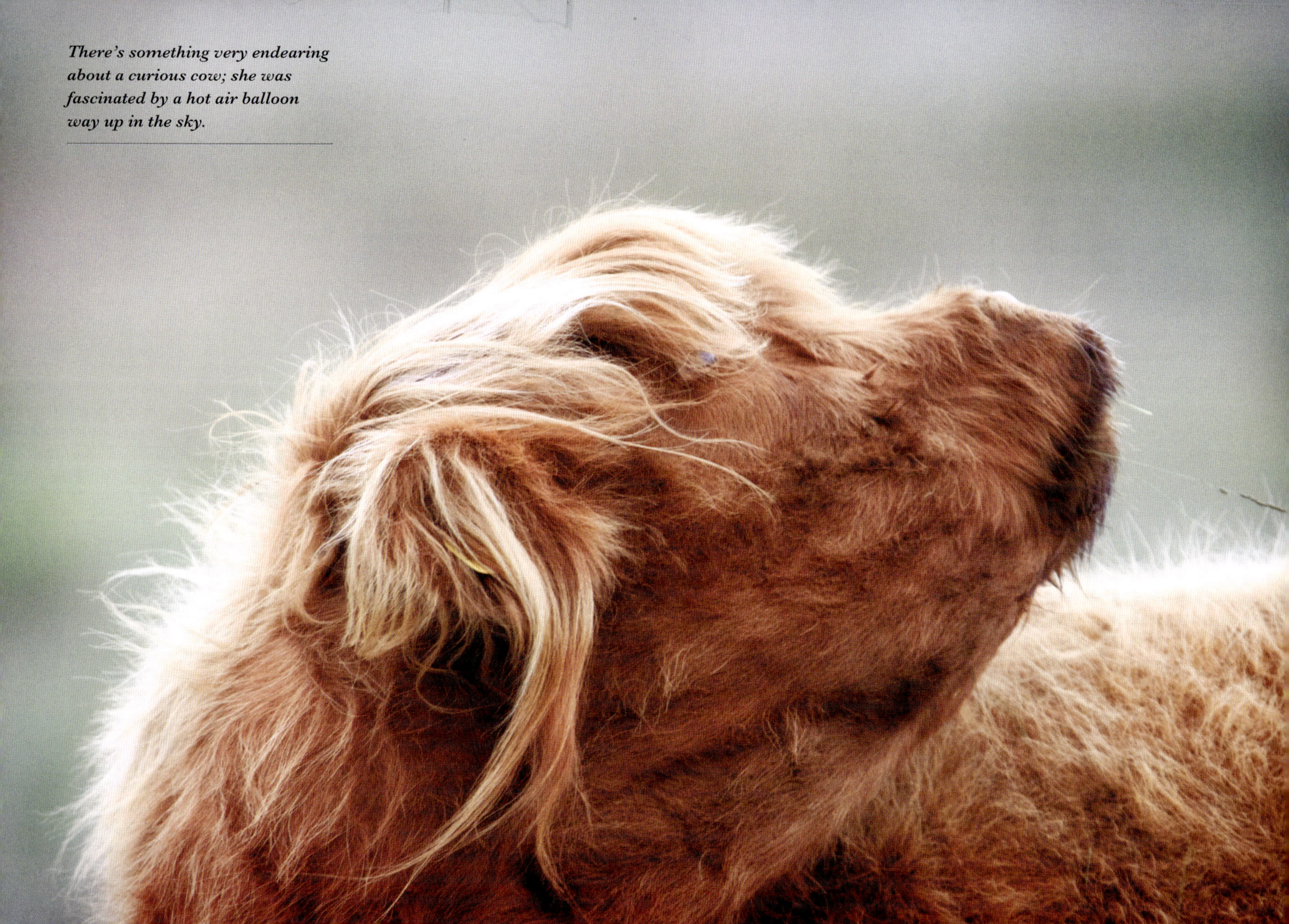
There's something very endearing
about a curious cow; she was
fascinated by a hot air balloon
way up in the sky.

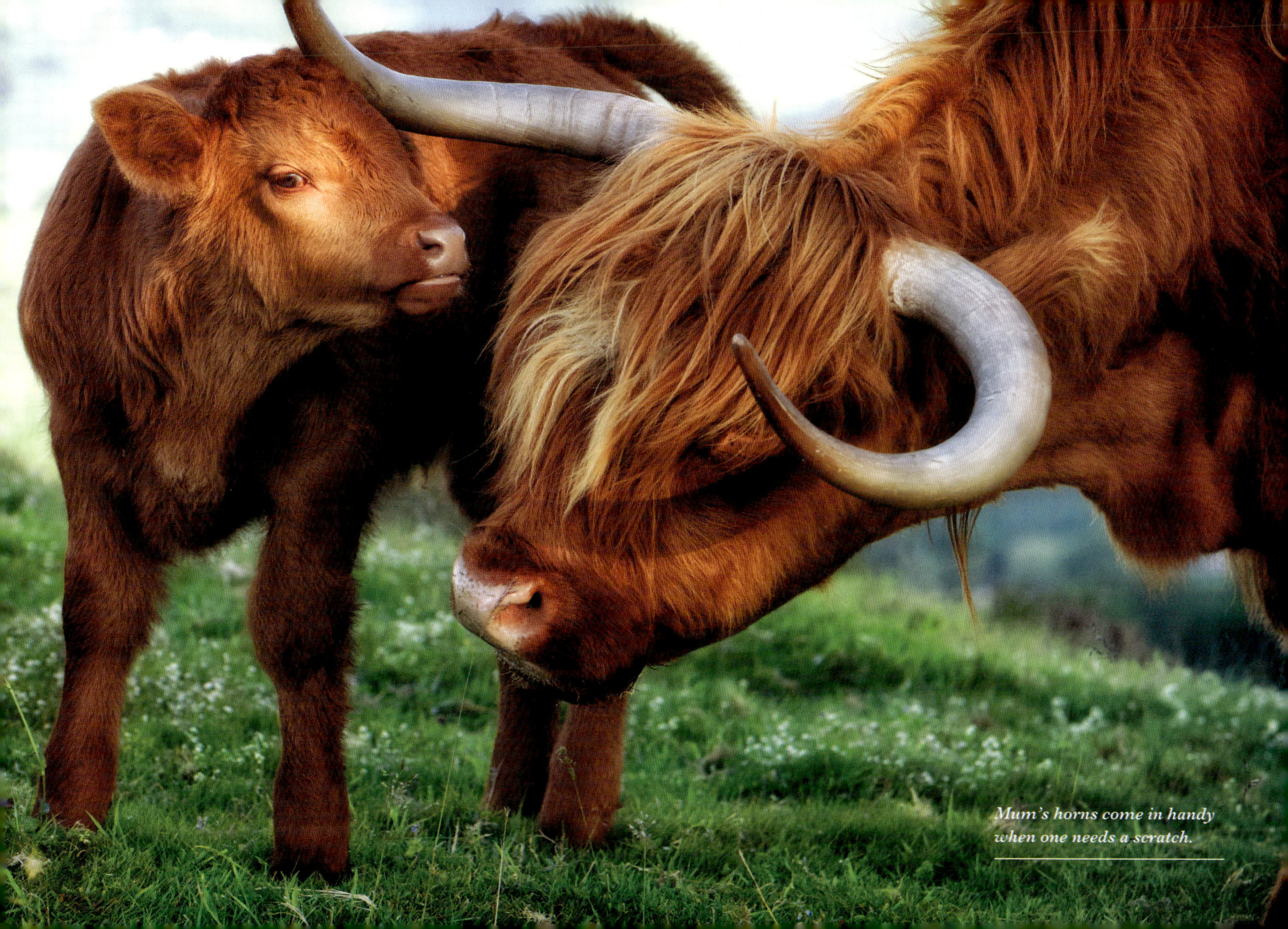

Mum's horns come in handy
when one needs a scratch.

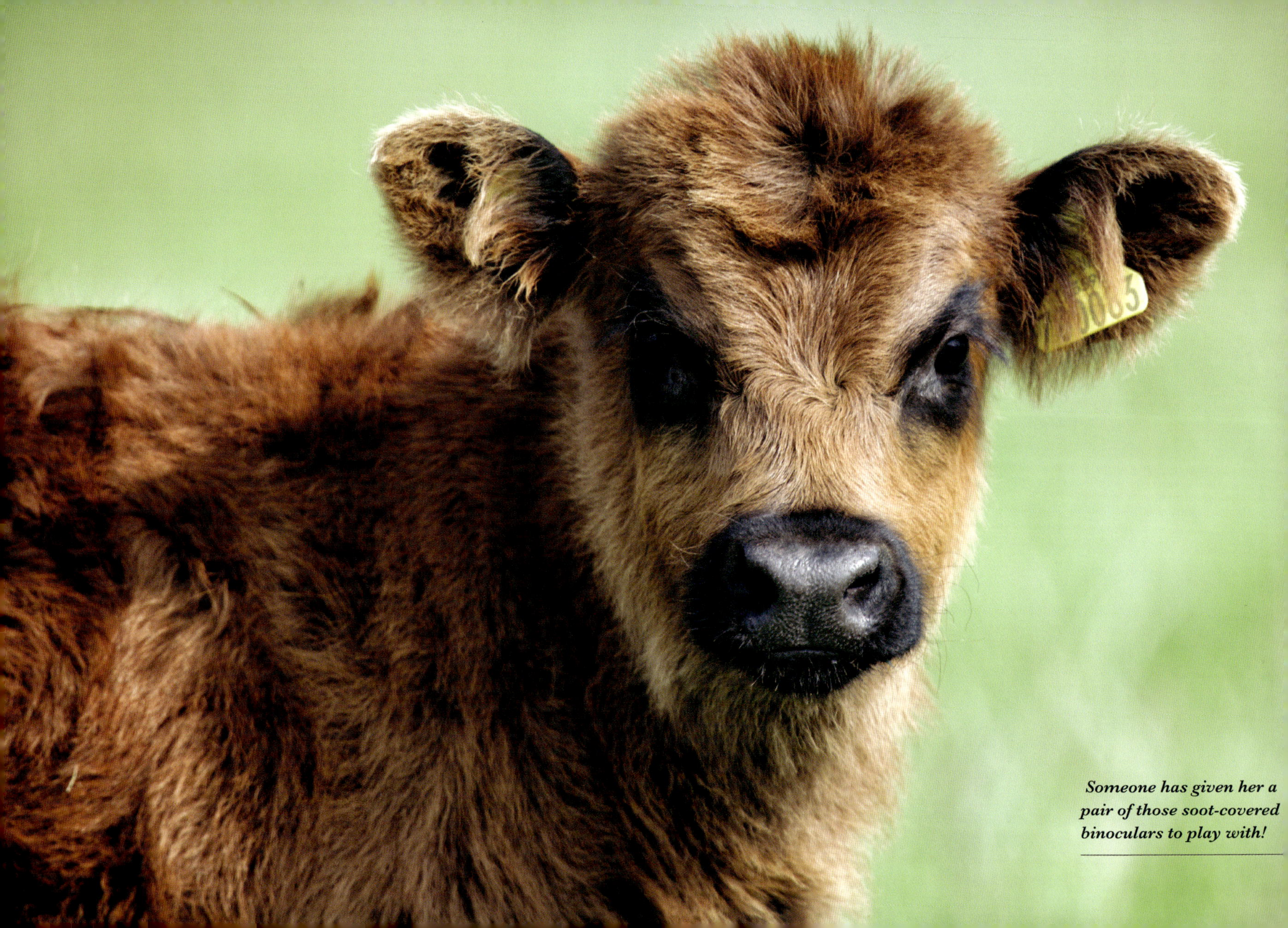

Someone has given her a pair of those soot-covered binoculars to play with!

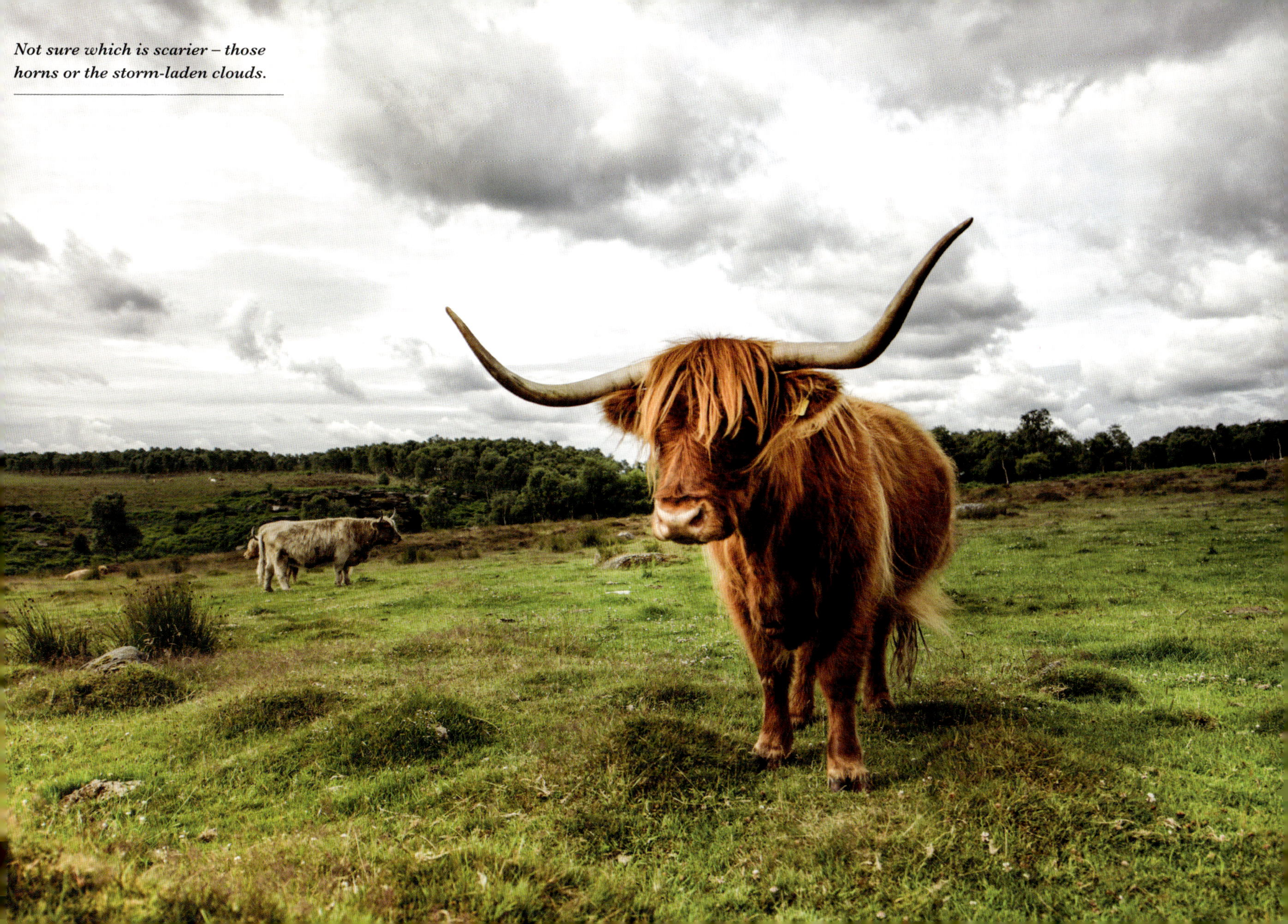
Not sure which is scarier – those
horns or the storm-laden clouds.

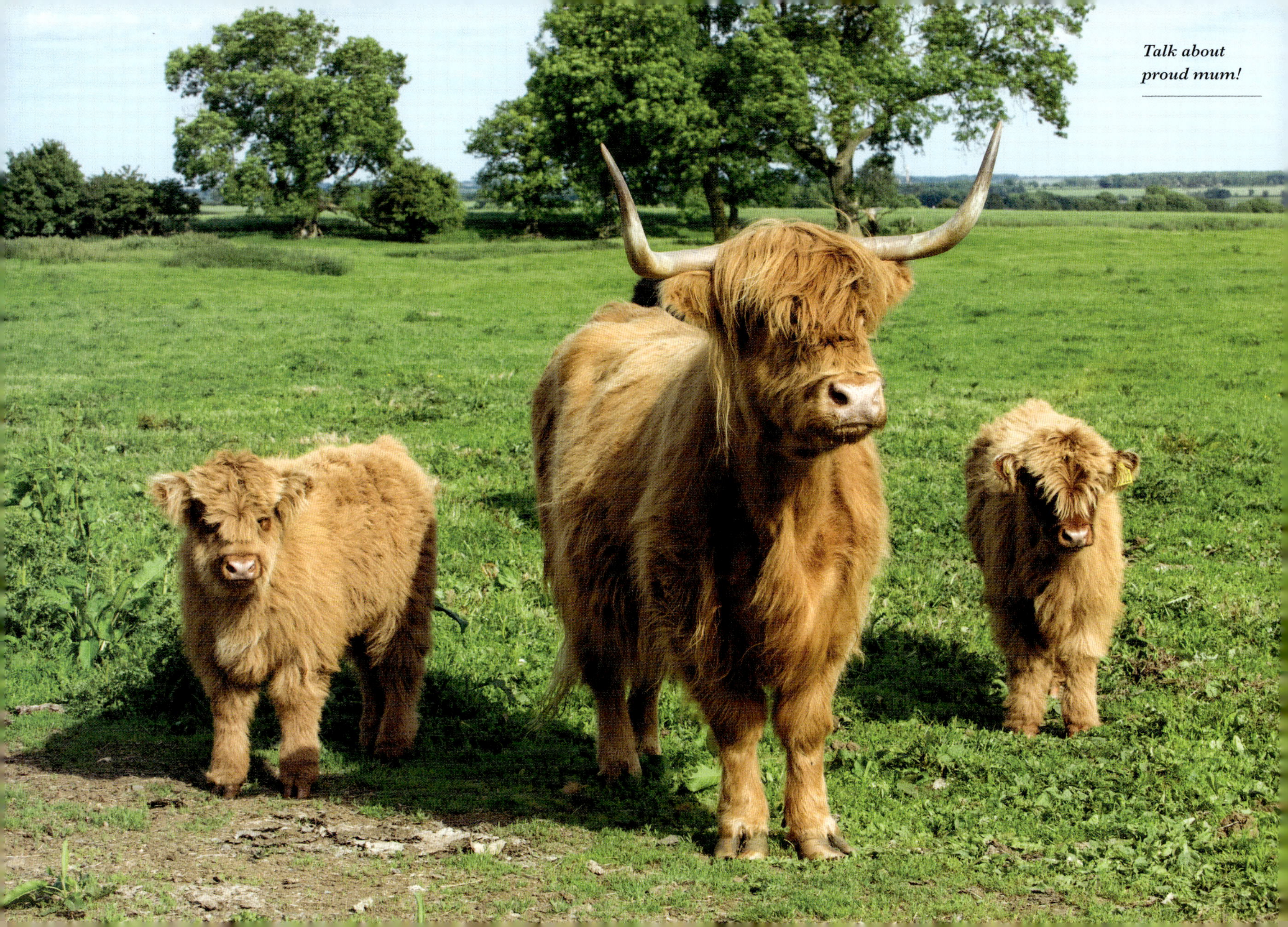
Talk about
proud mum!

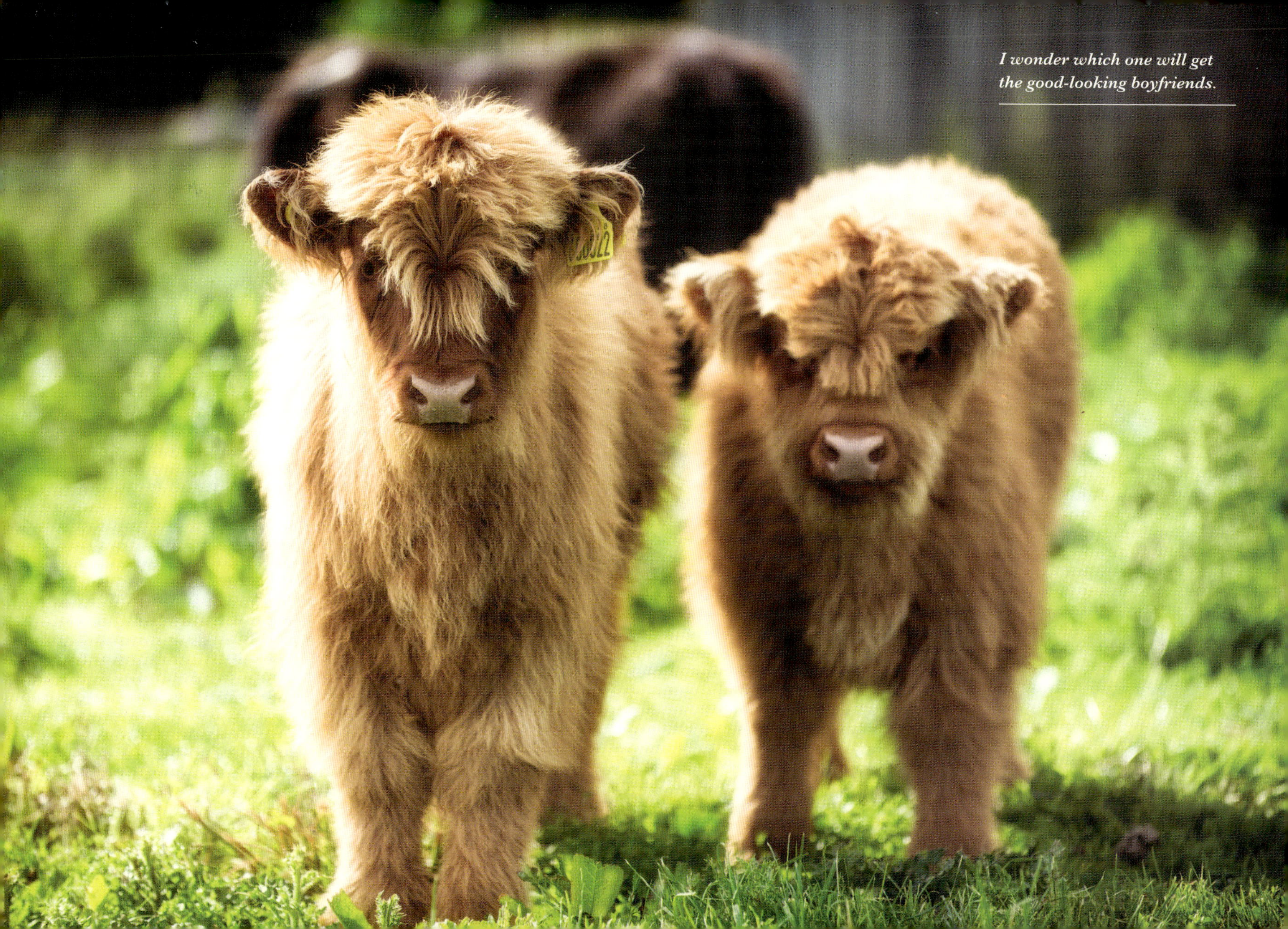

I wonder which one will get
the good-looking boyfriends.

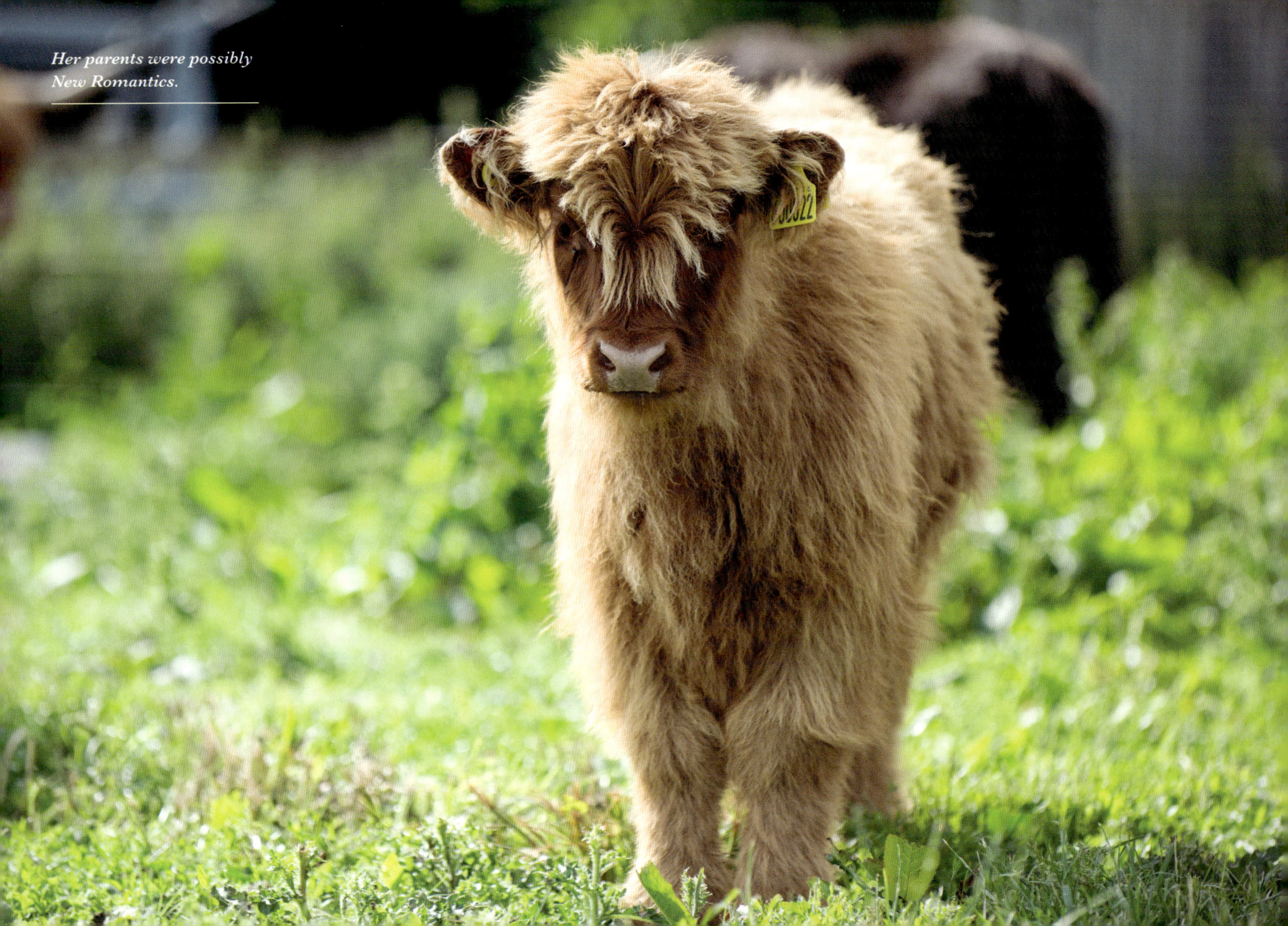
Her parents were possibly
New Romantics.

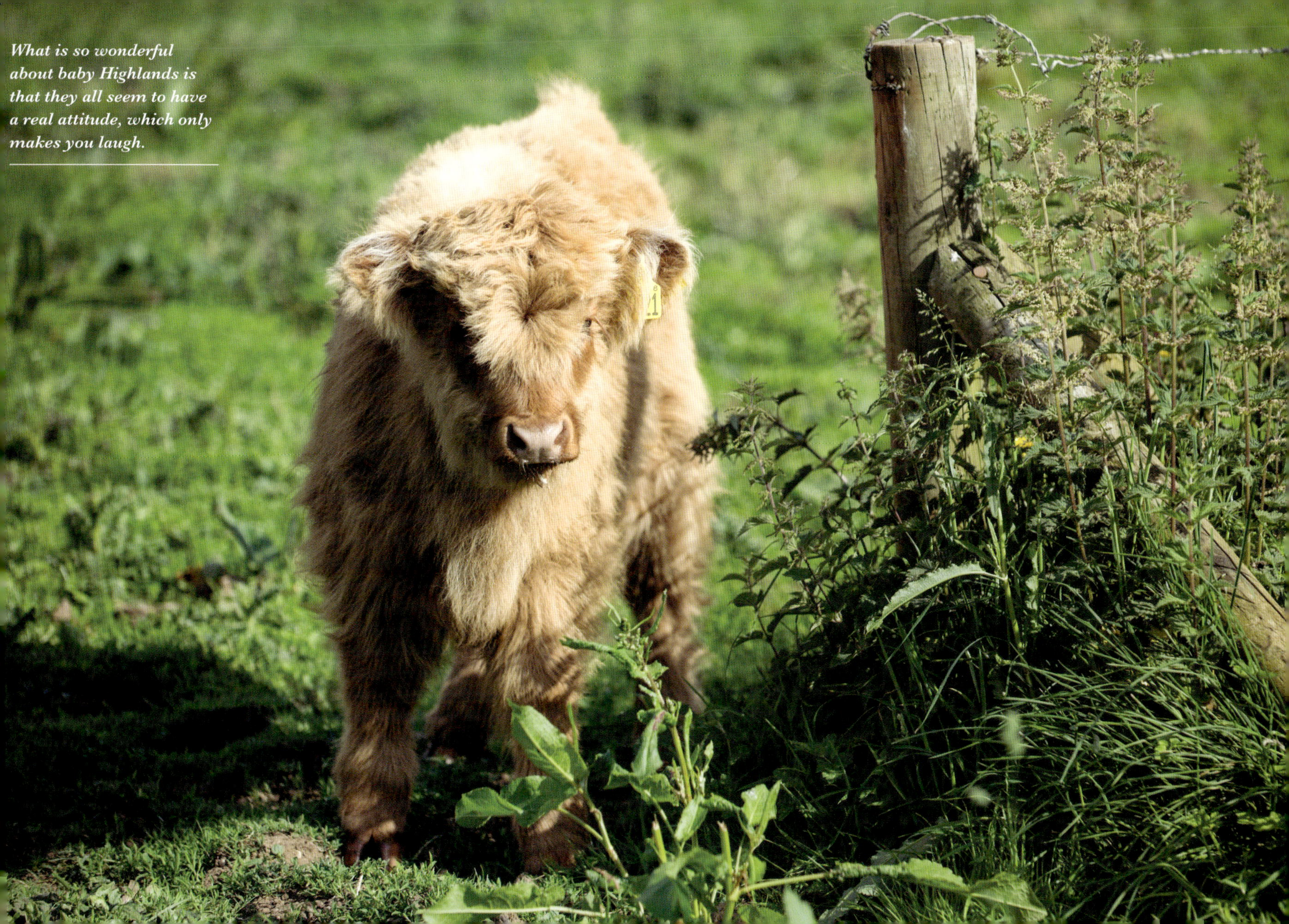
What is so wonderful
about baby Highlands is
that they all seem to have
a real attitude, which only
makes you laugh.

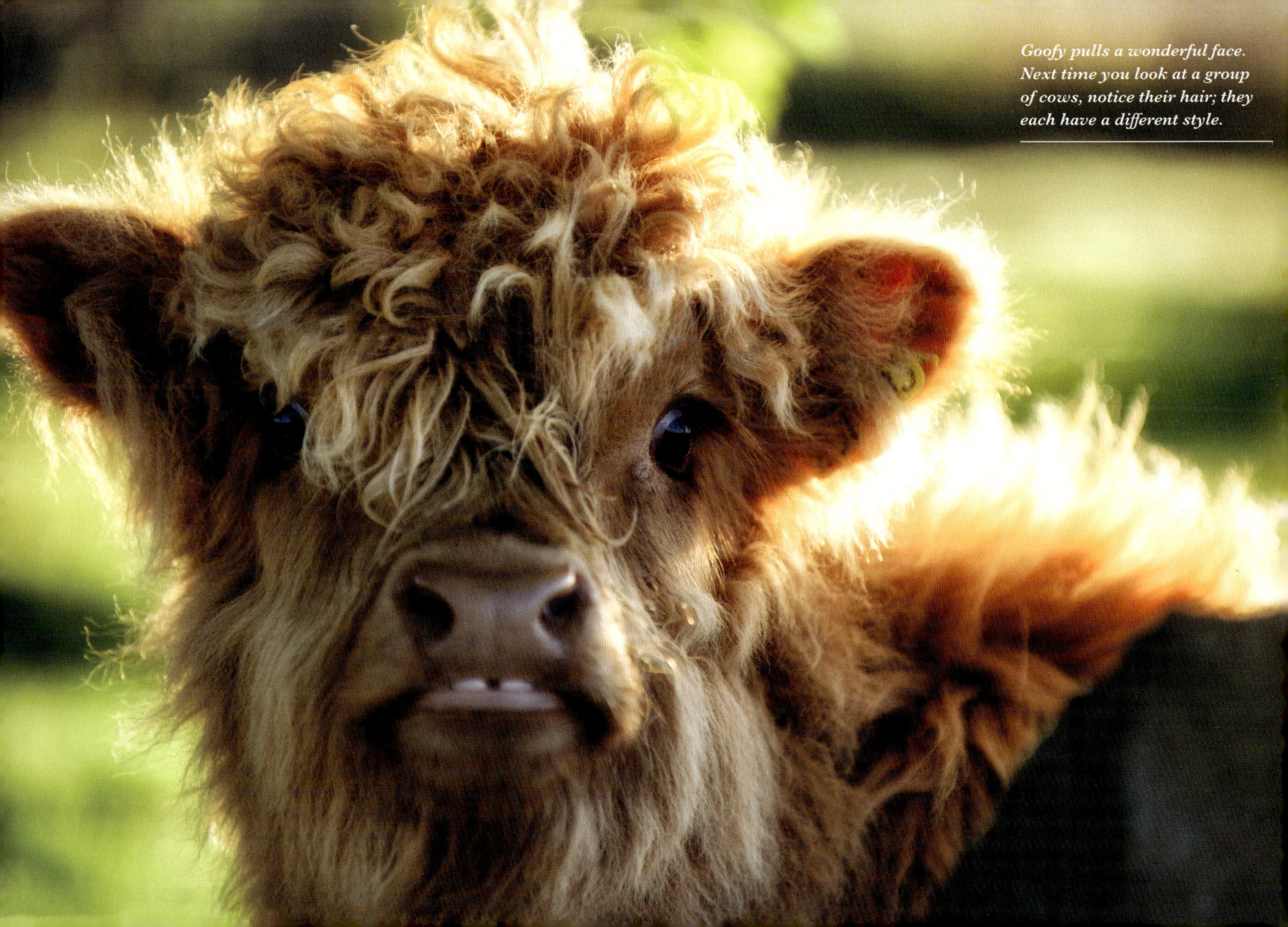

Goofy pulls a wonderful face.
Next time you look at a group
of cows, notice their hair; they
each have a different style.

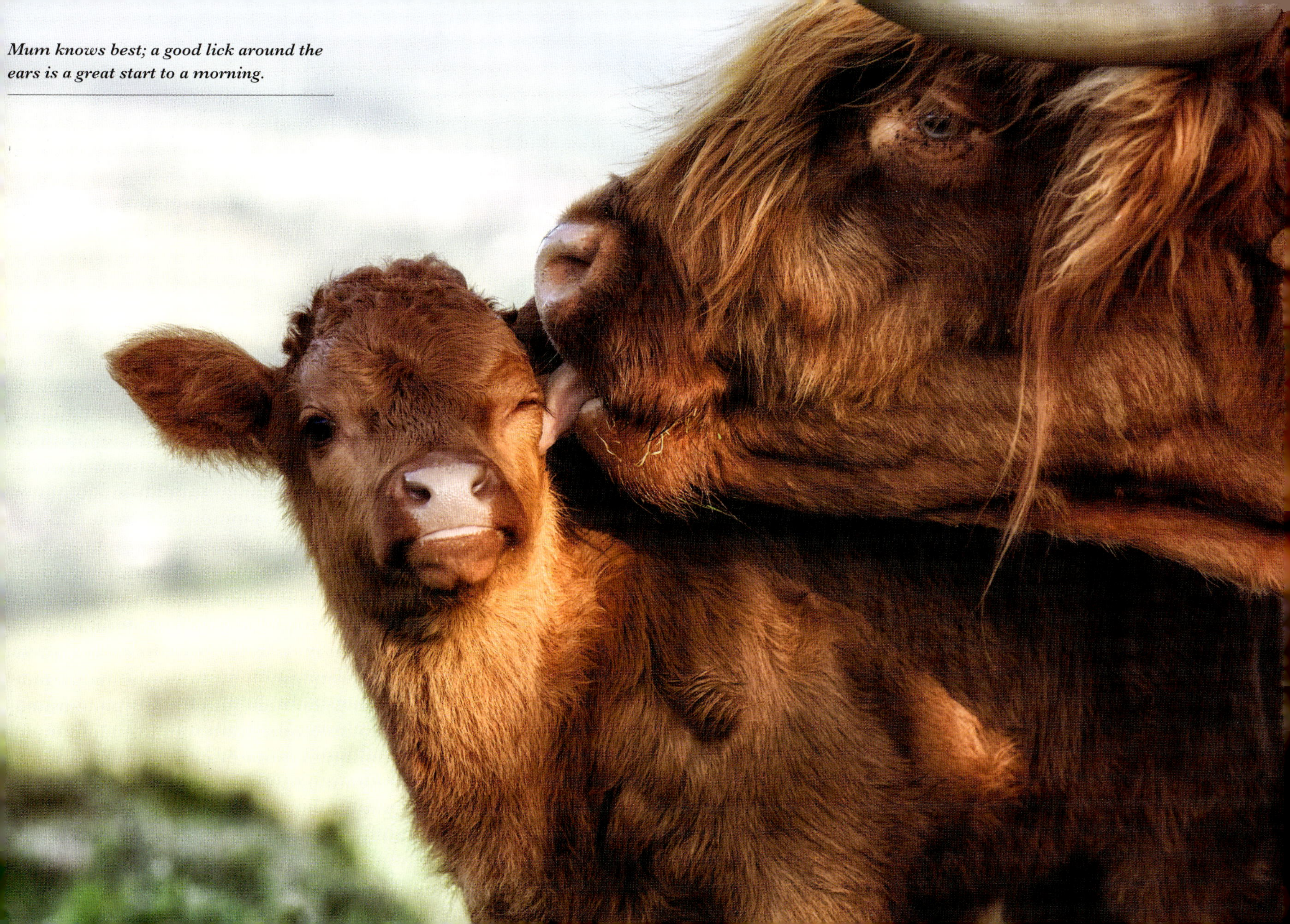

Mum knows best; a good lick around the ears is a great start to a morning.

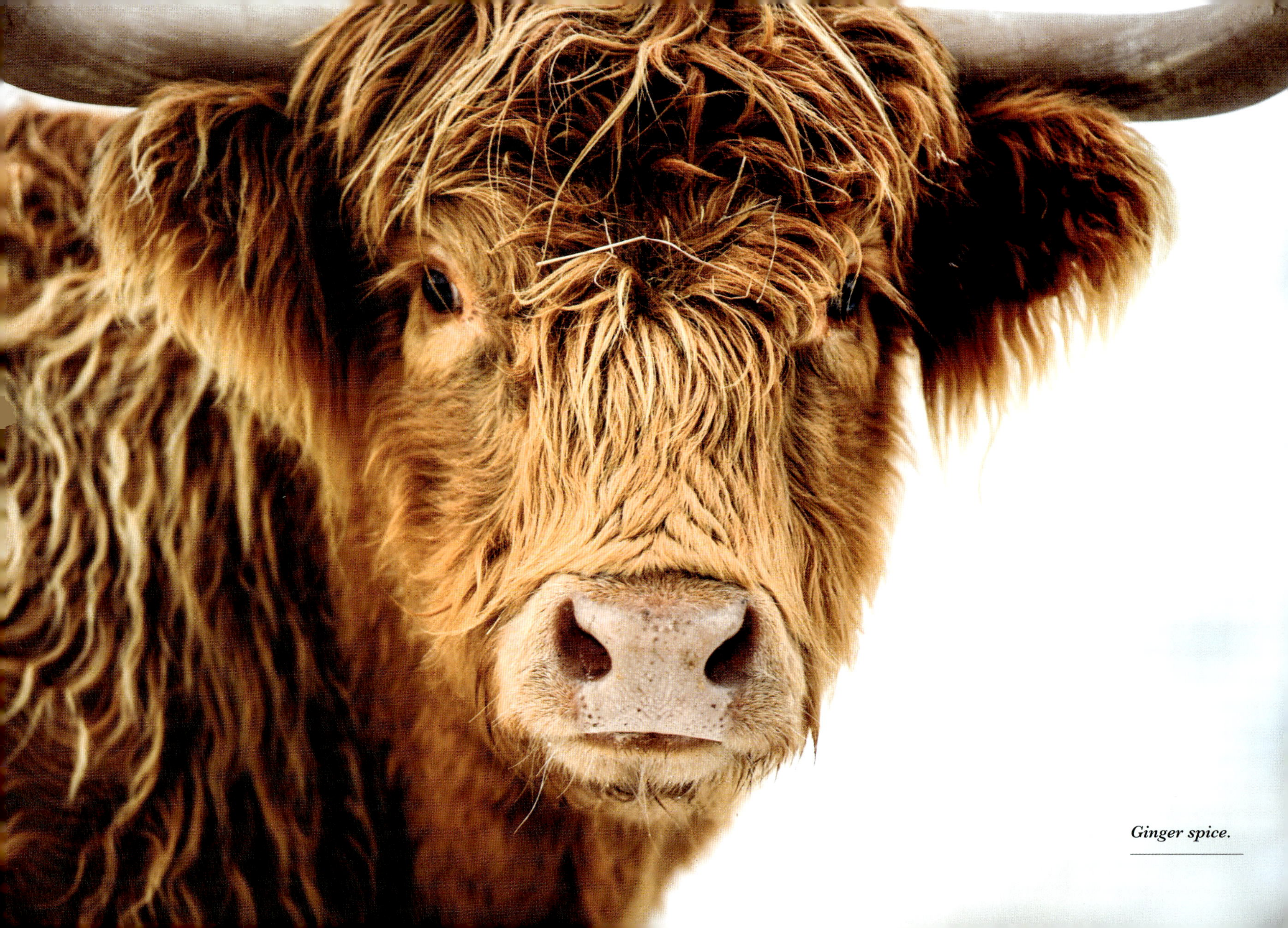

Ginger spice.

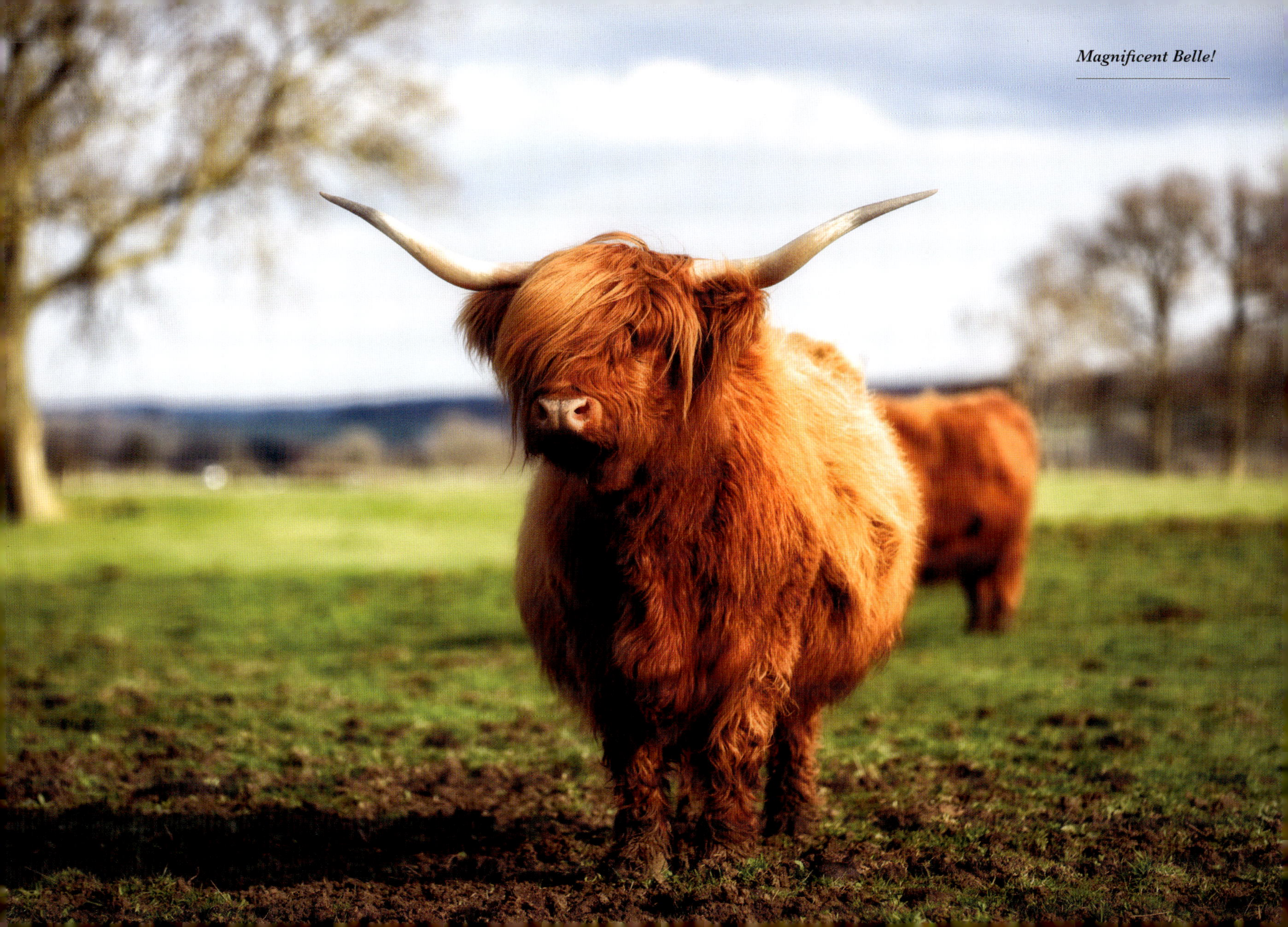
Magnificent Belle!

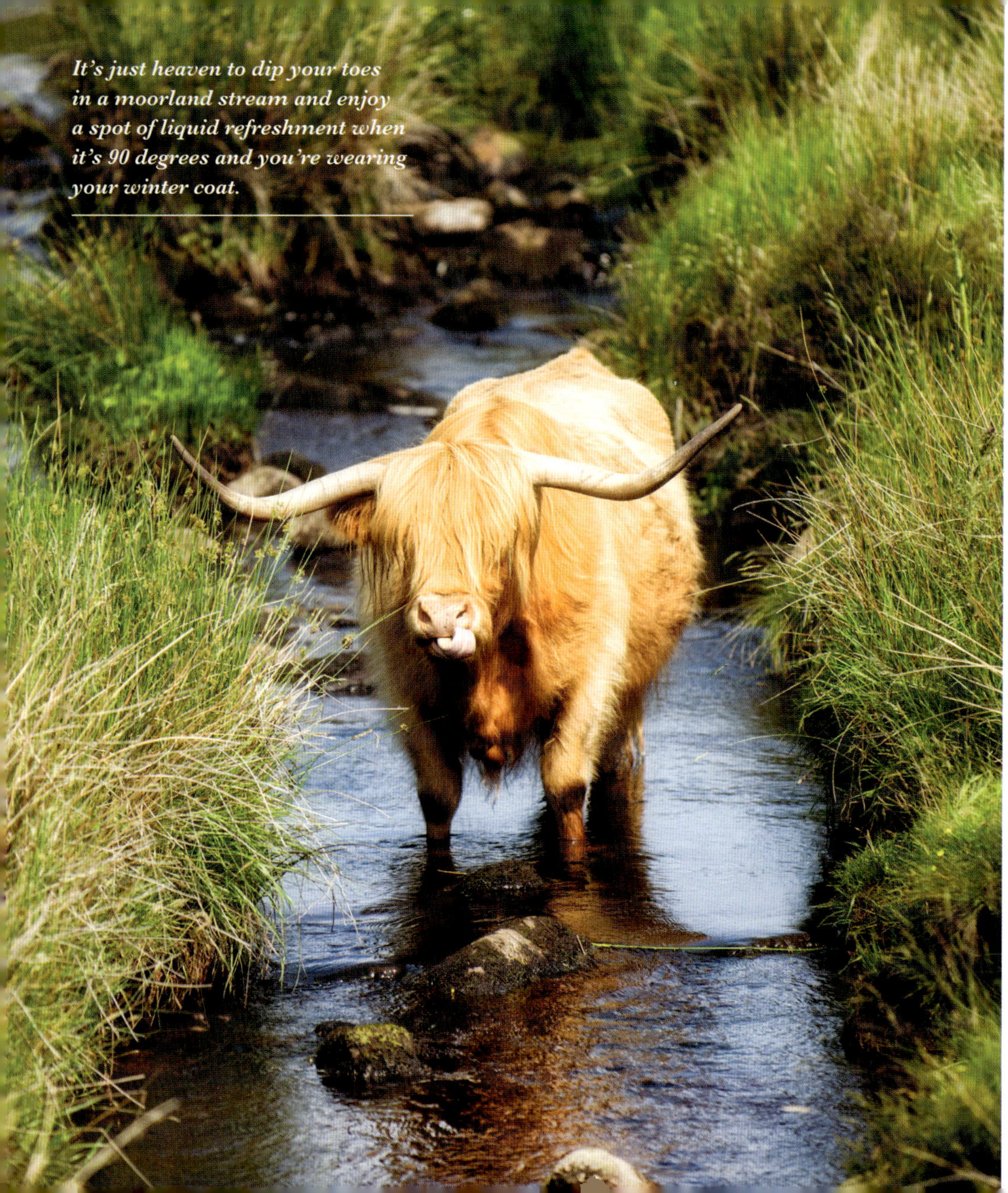

It's just heaven to dip your toes
in a moorland stream and enjoy
a spot of liquid refreshment when
it's 90 degrees and you're wearing
your winter coat.

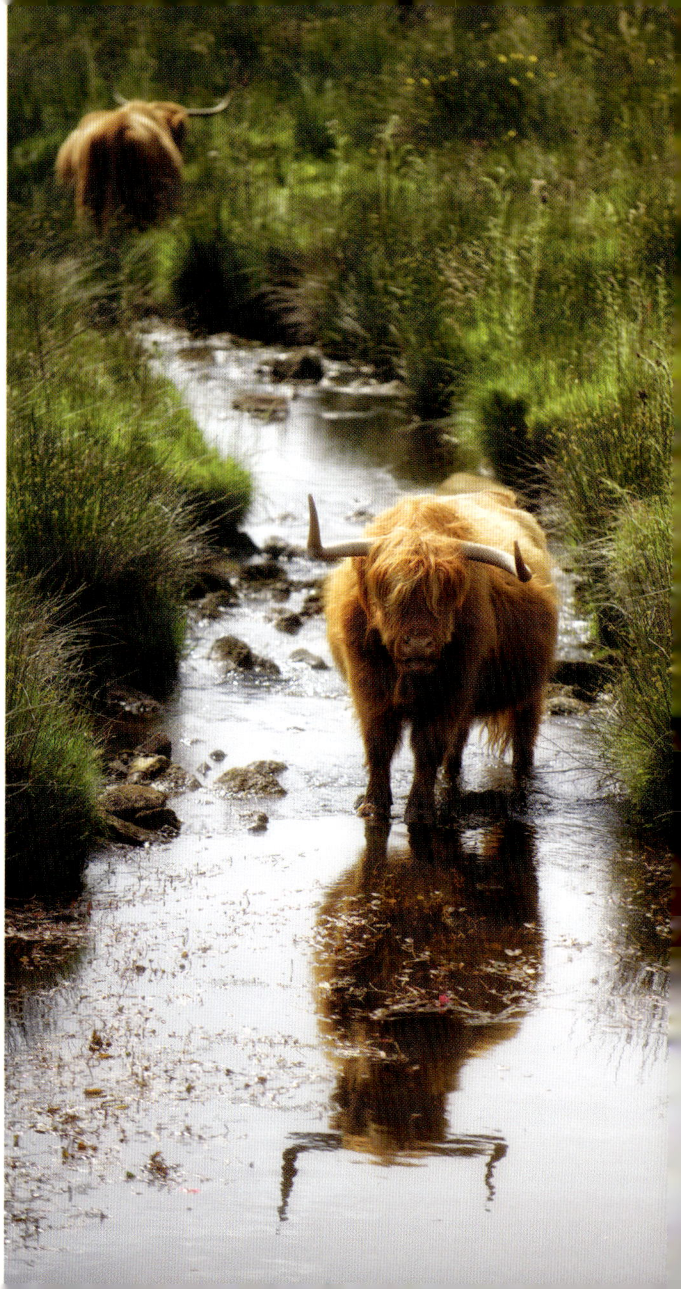

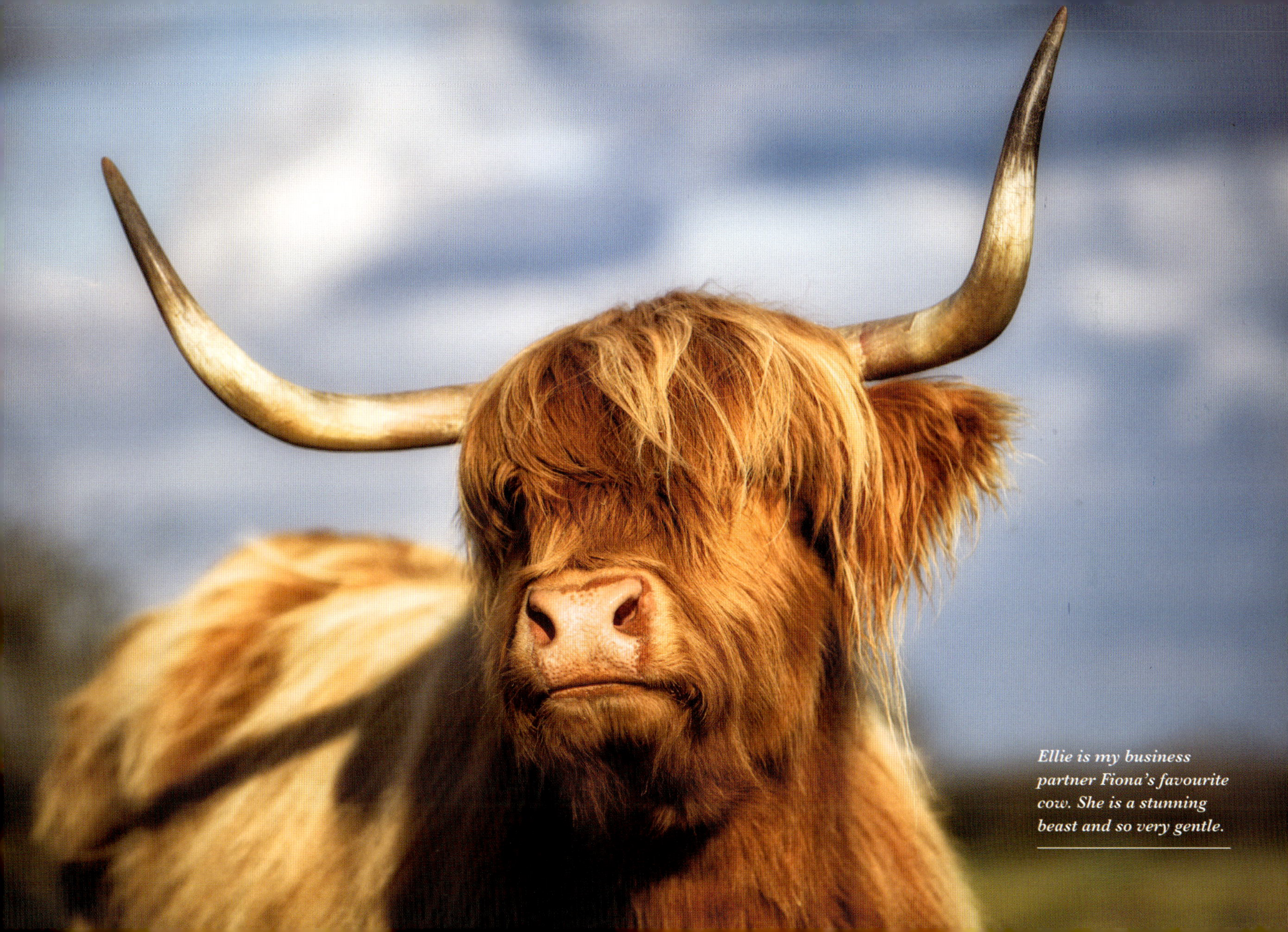

Ellie is my business partner Fiona's favourite cow. She is a stunning beast and so very gentle.

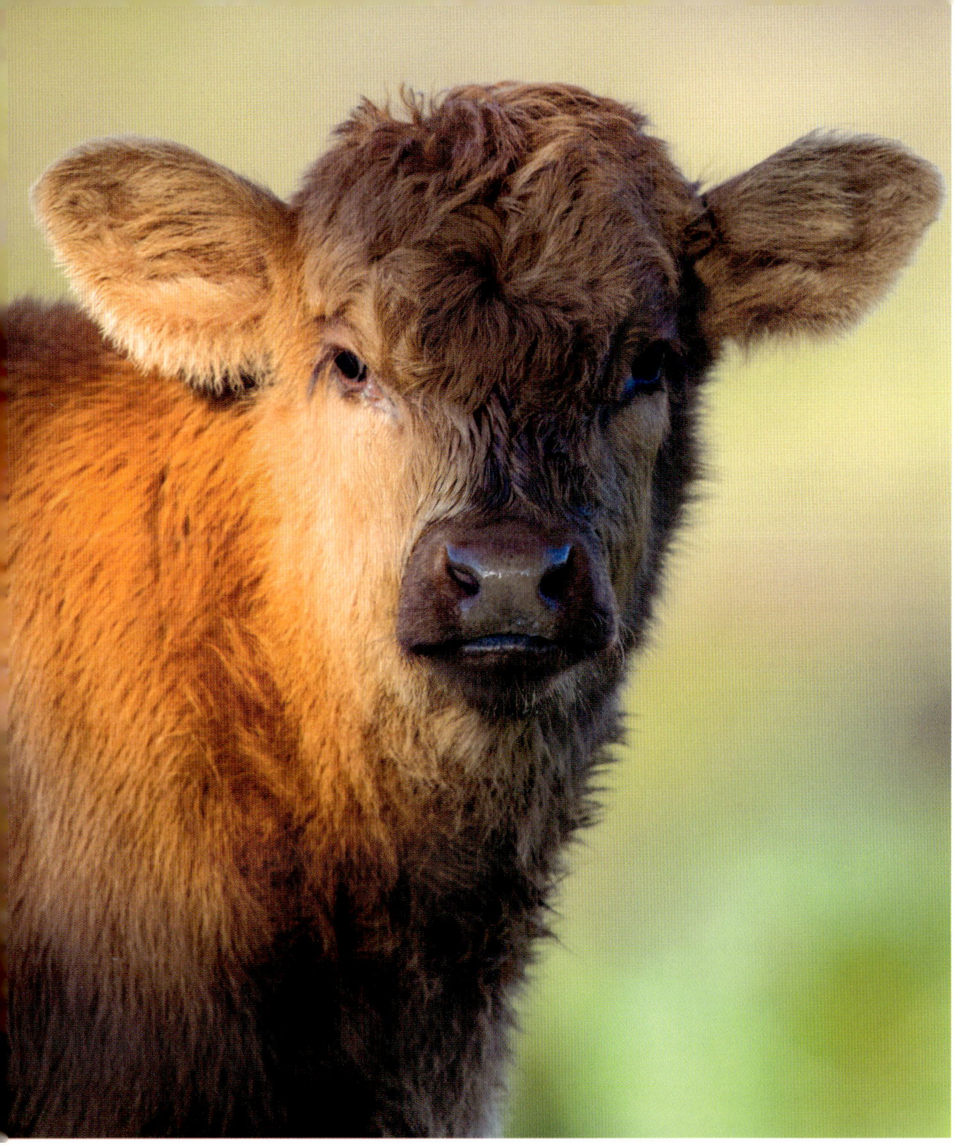

Wet nose.

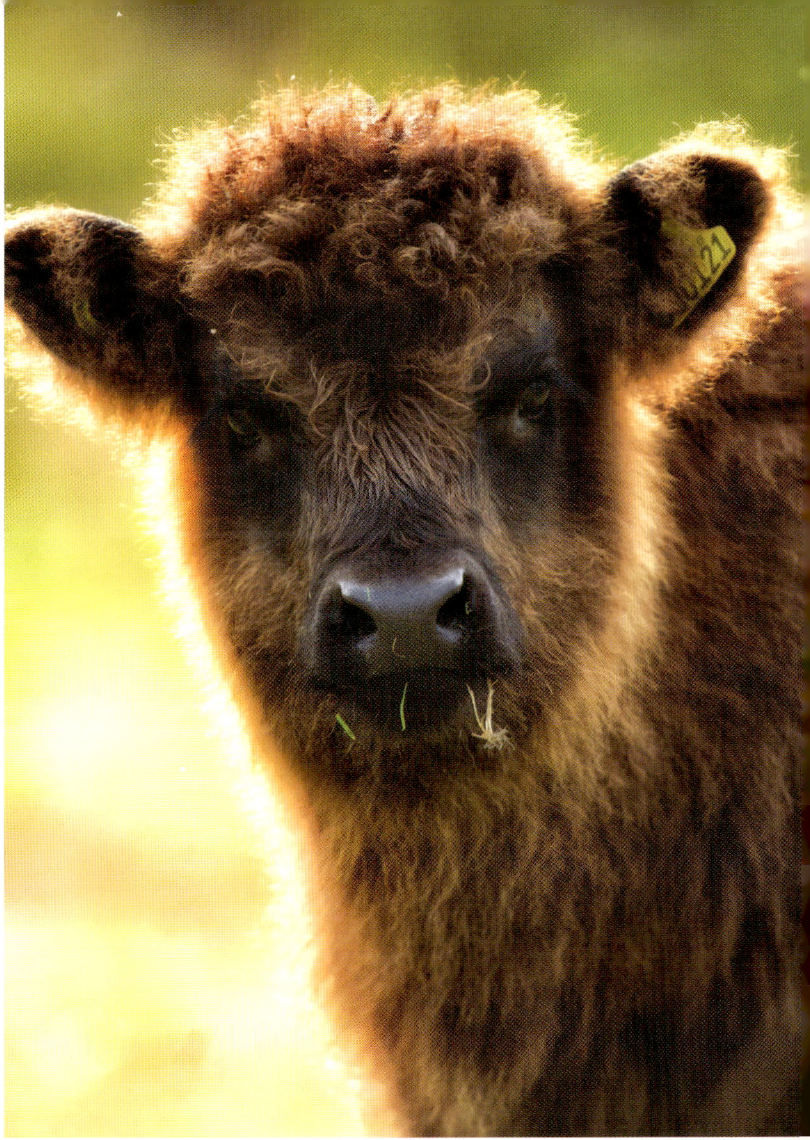

Mid-breakfast.

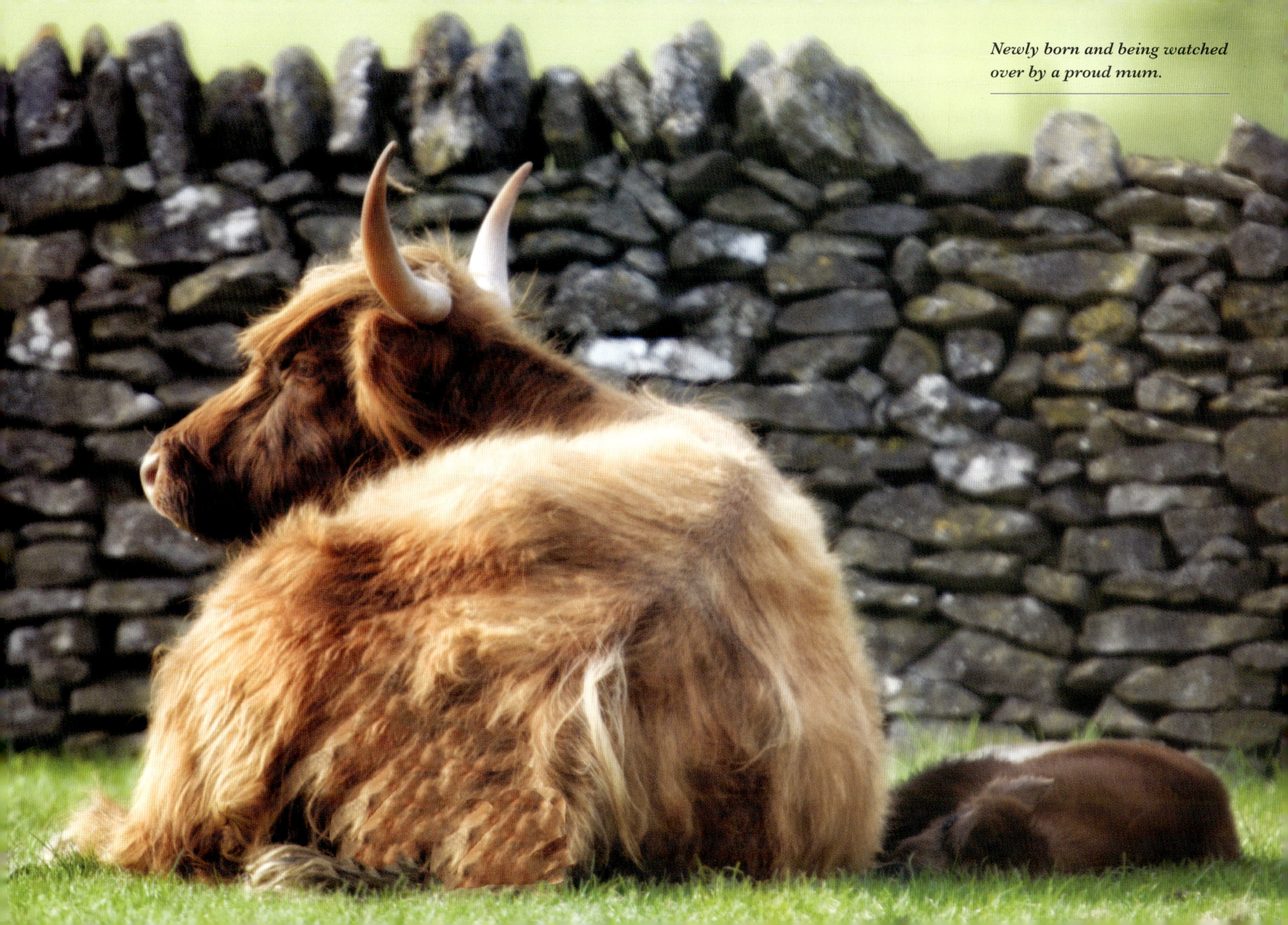

Newly born and being watched
over by a proud mum.

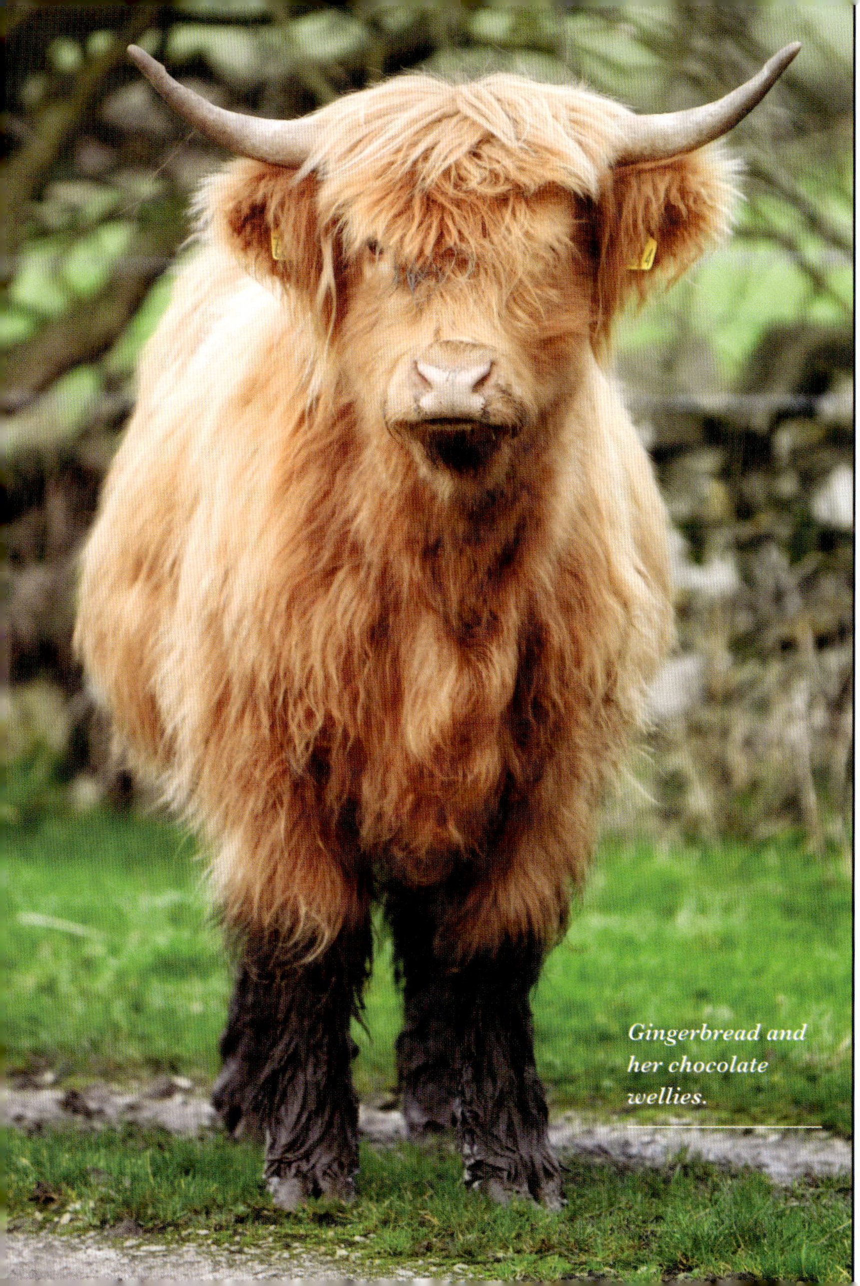

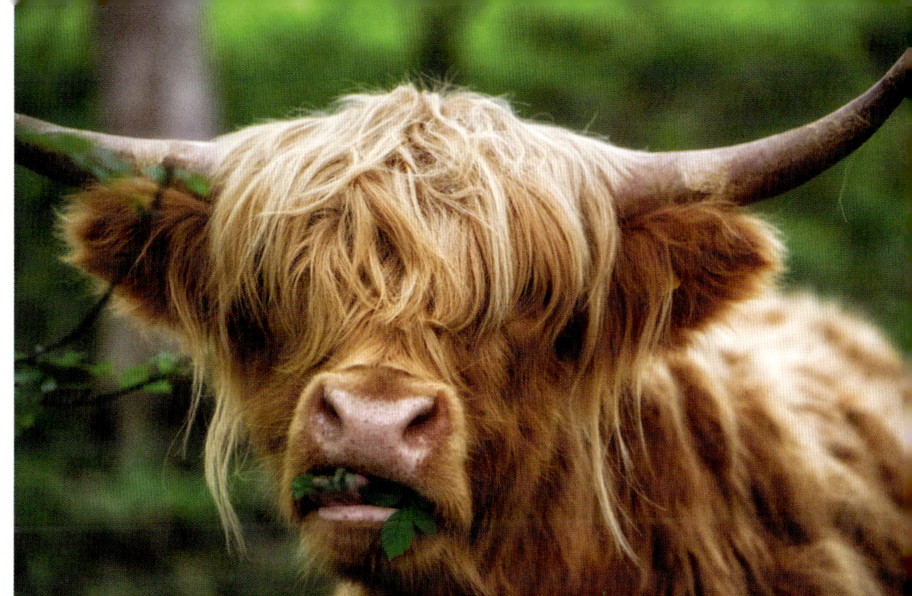

I asked Trixibelle as politely as I could to move away from the branch so I could get a clear shot, and this was the response. Cheeky blooming moo!

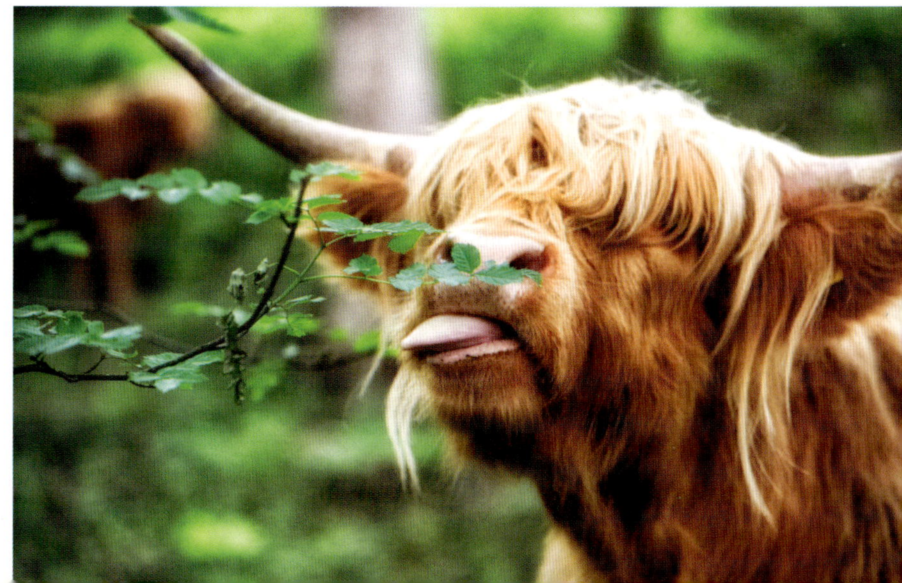

Gingerbread and her chocolate wellies.

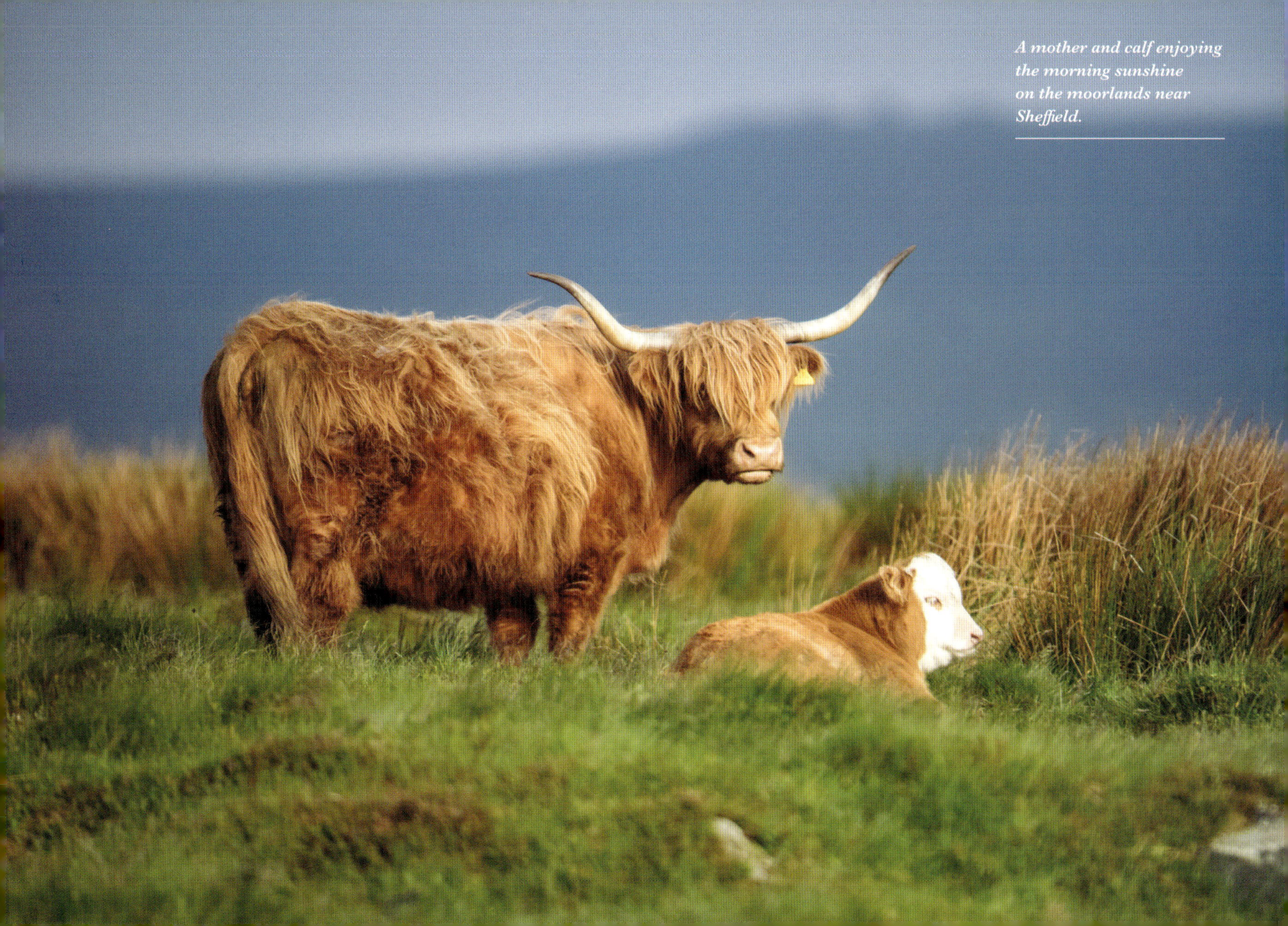

A mother and calf enjoying
the morning sunshine
on the moorlands near
Sheffield.

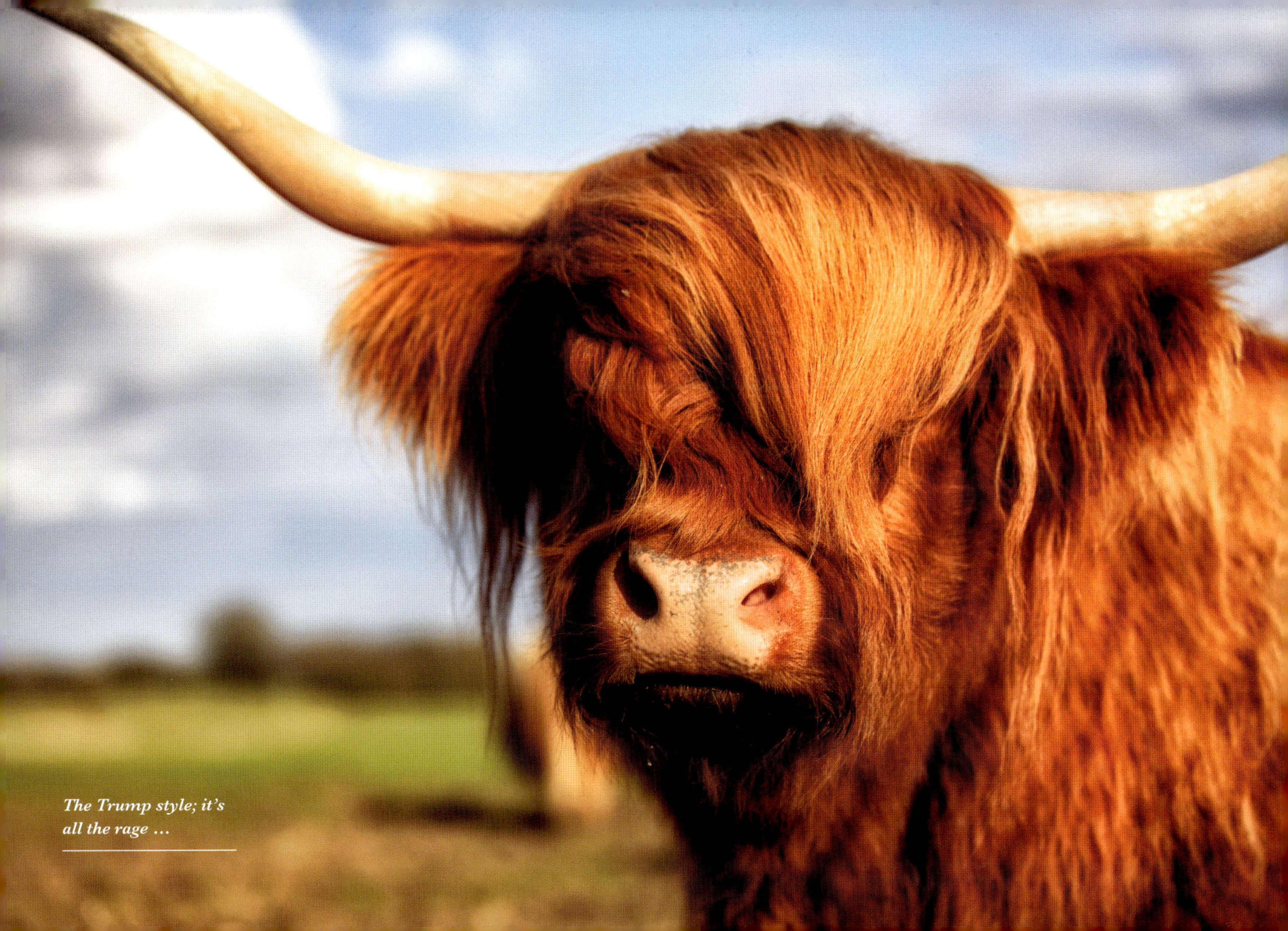

The Trump style; it's
all the rage …

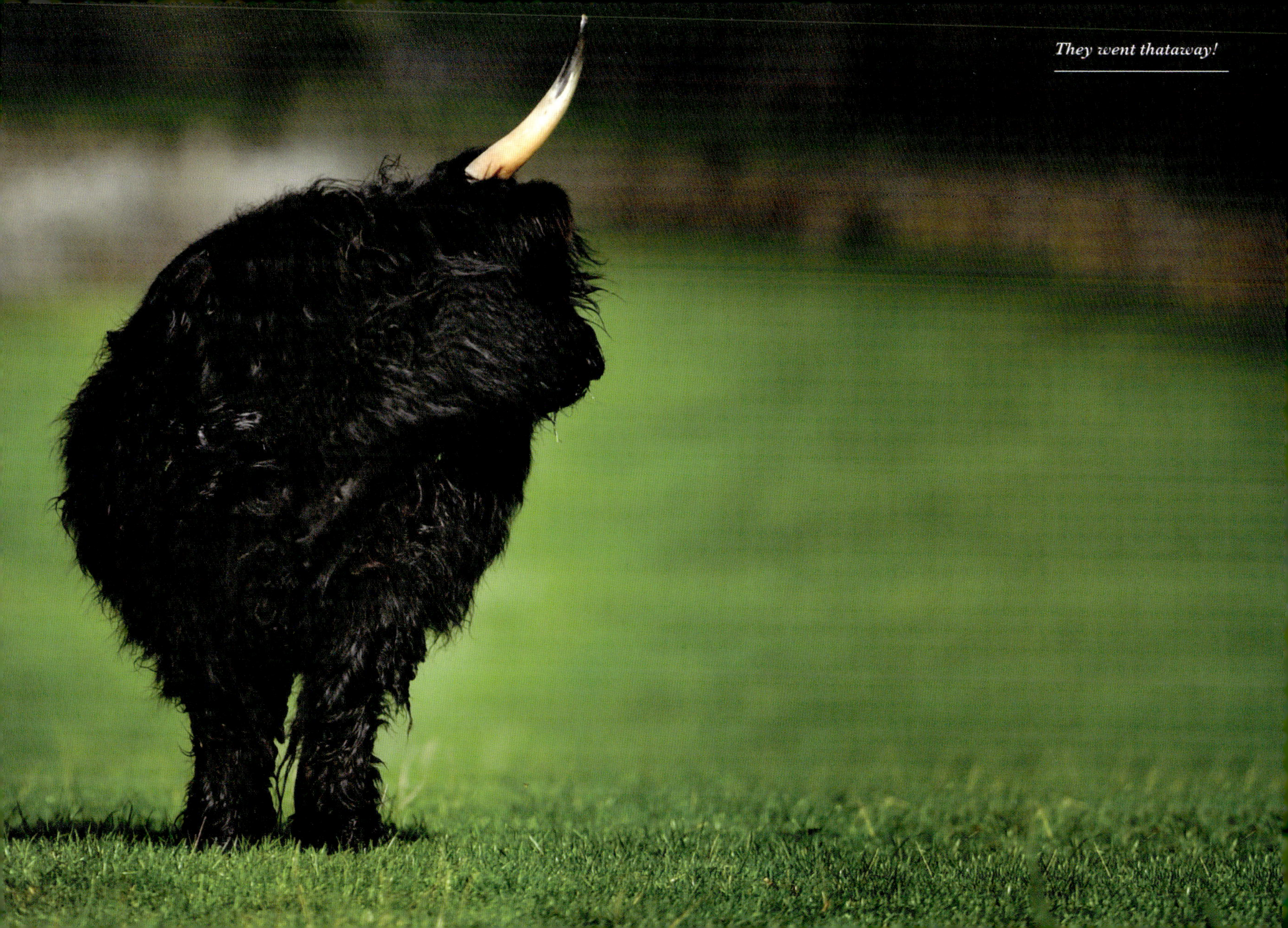
They went thataway!

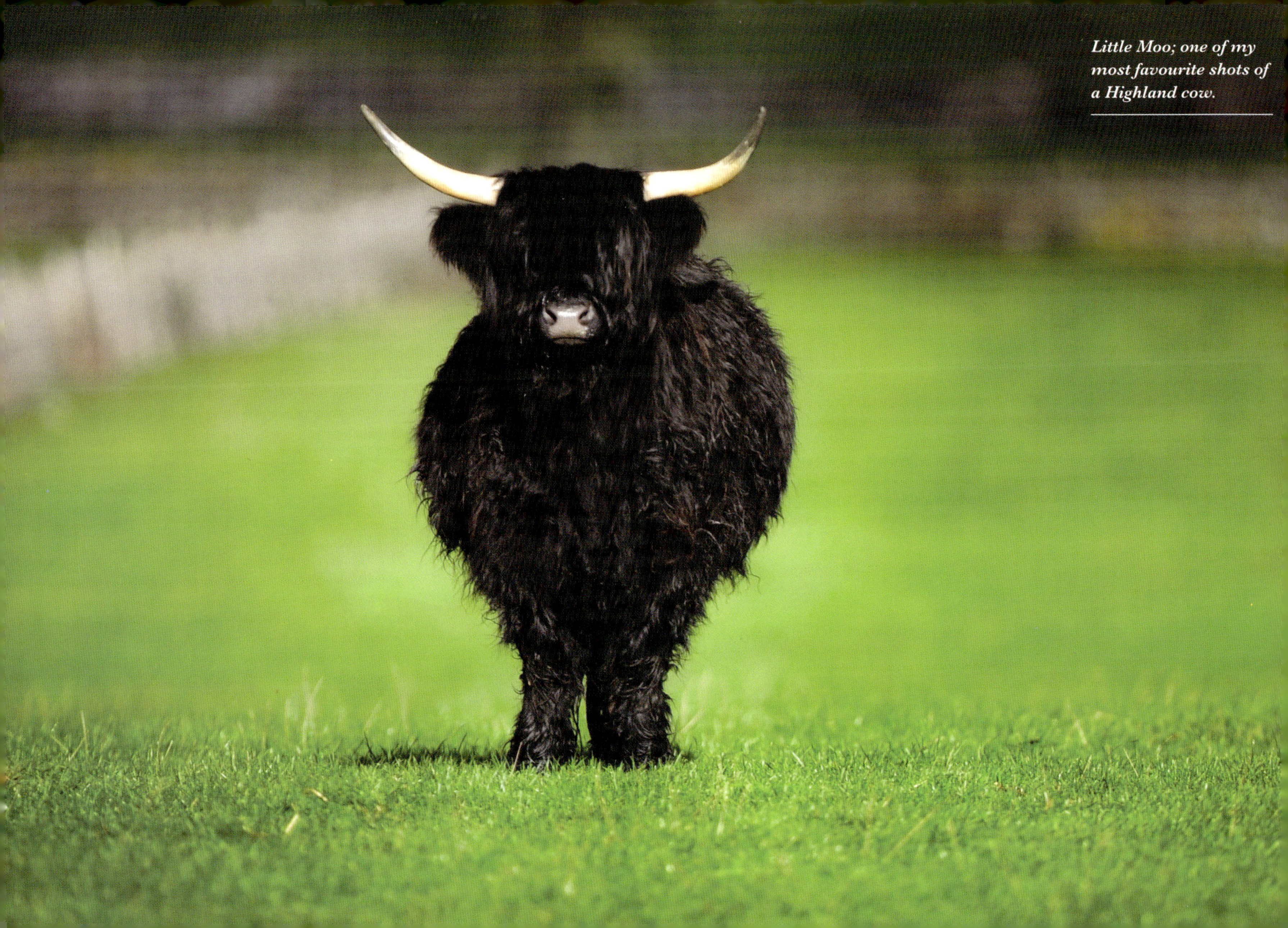

Little Moo; one of my
most favourite shots of
a Highland cow.

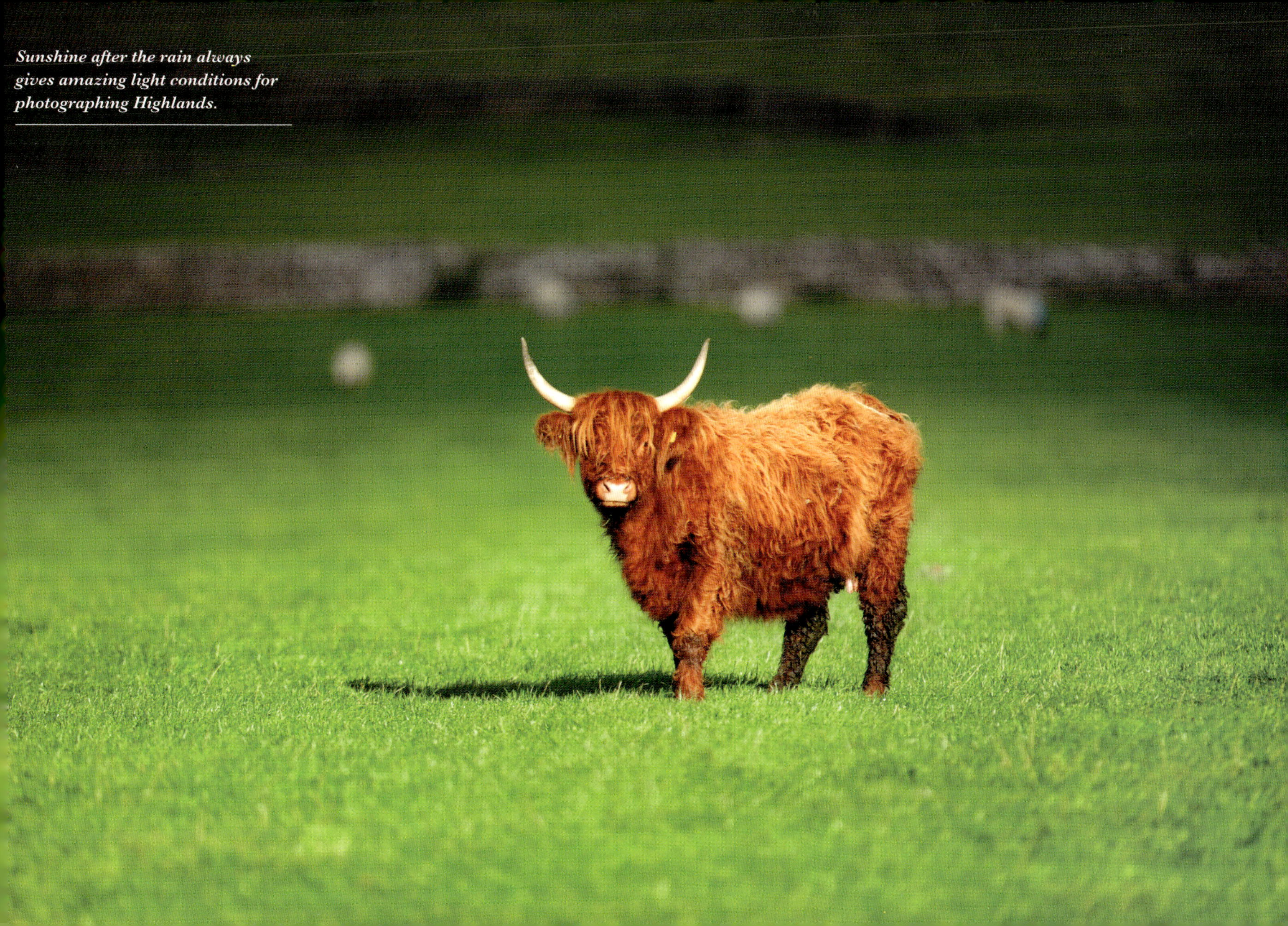
Sunshine after the rain always
gives amazing light conditions for
photographing Highlands.

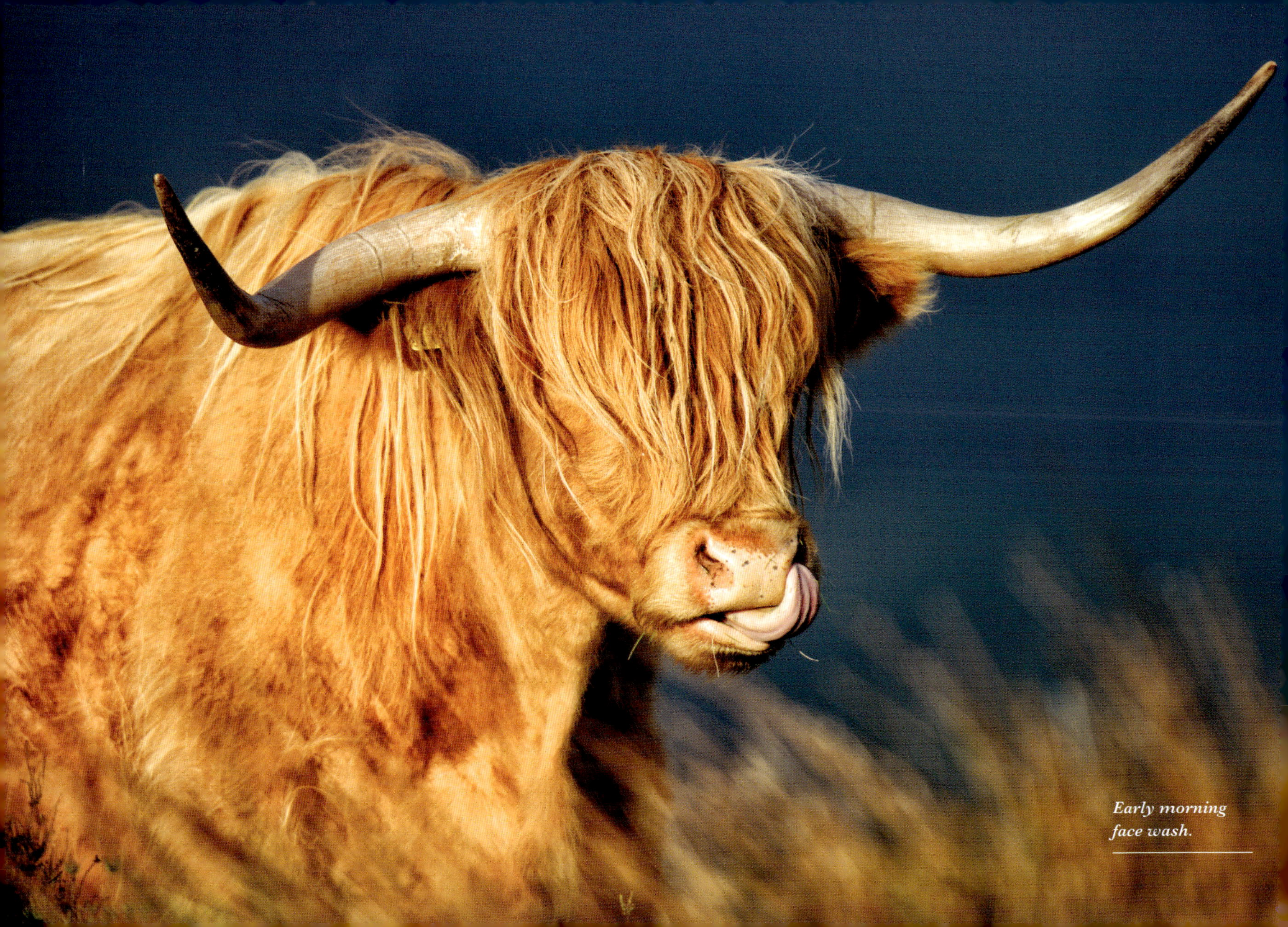

Early morning
face wash.

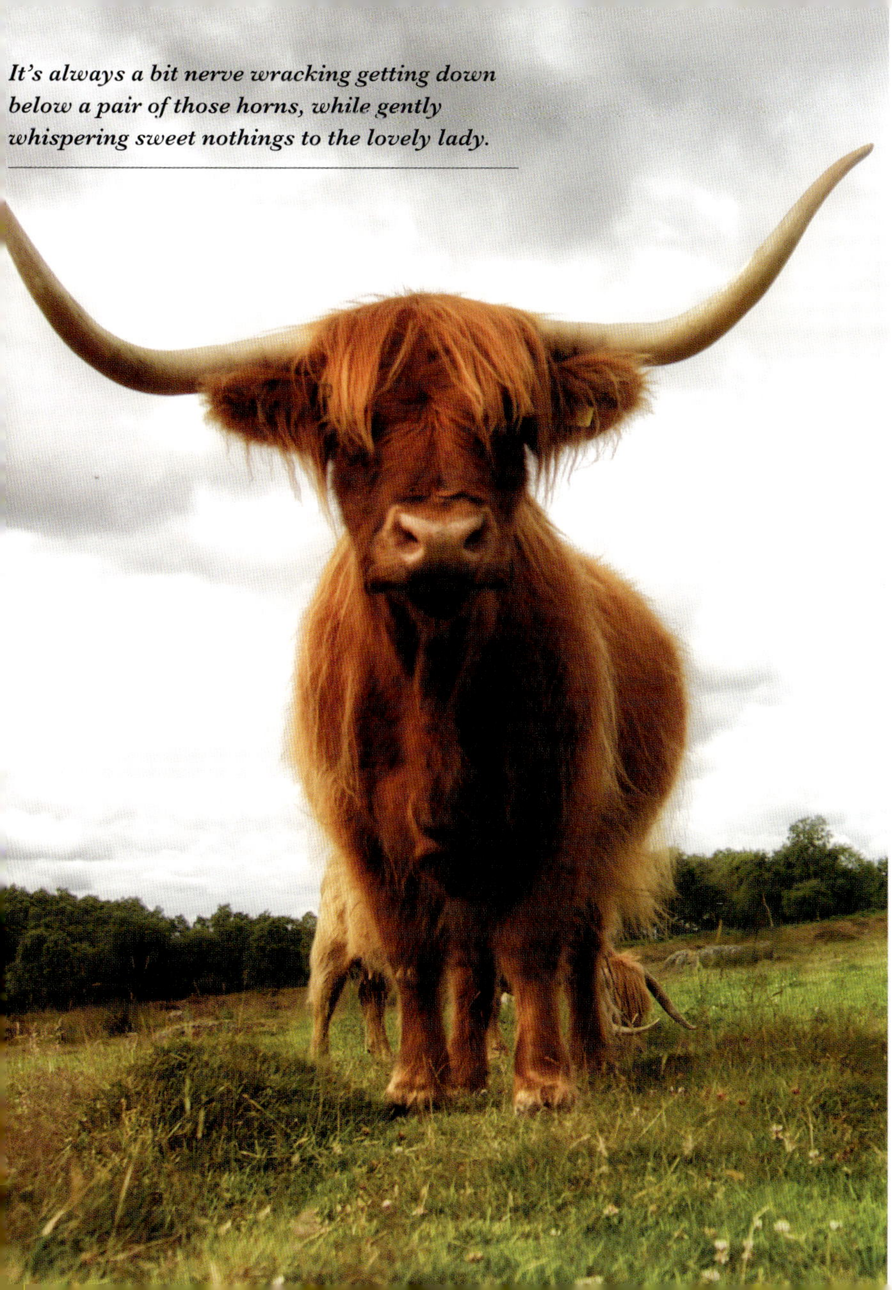

It's always a bit nerve wracking getting down below a pair of those horns, while gently whispering sweet nothings to the lovely lady.

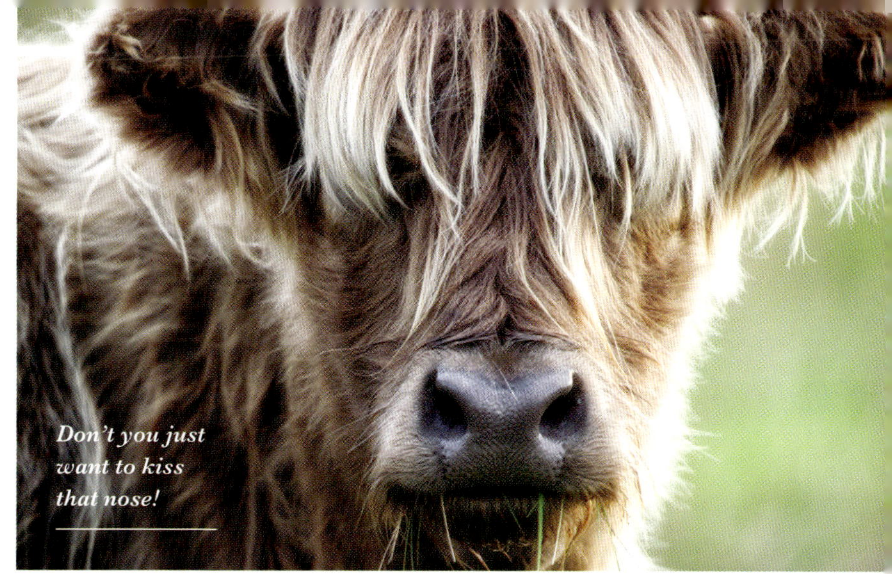
Don't you just want to kiss that nose!

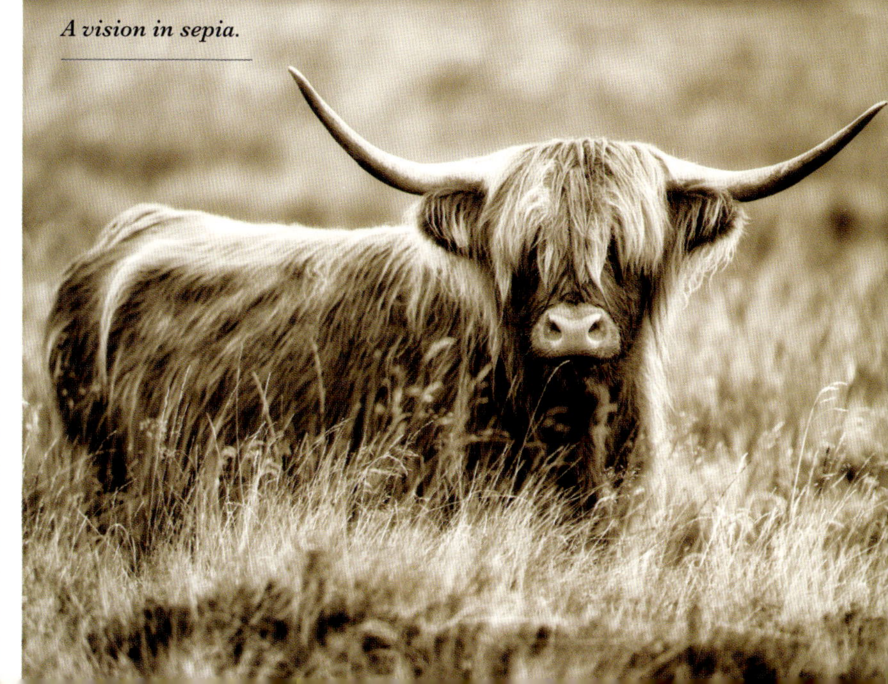
A vision in sepia.

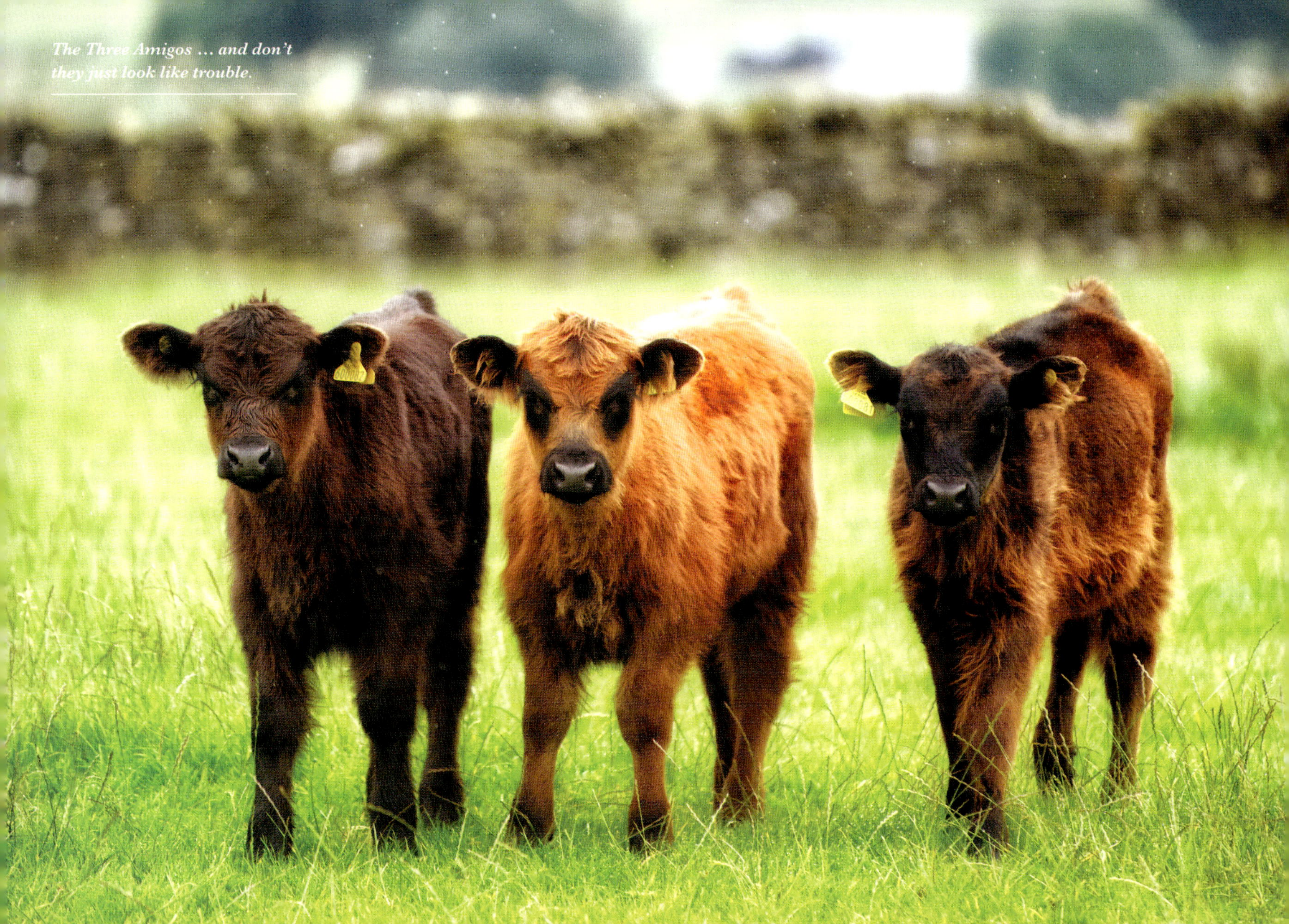

The Three Amigos … and don't
they just look like trouble.

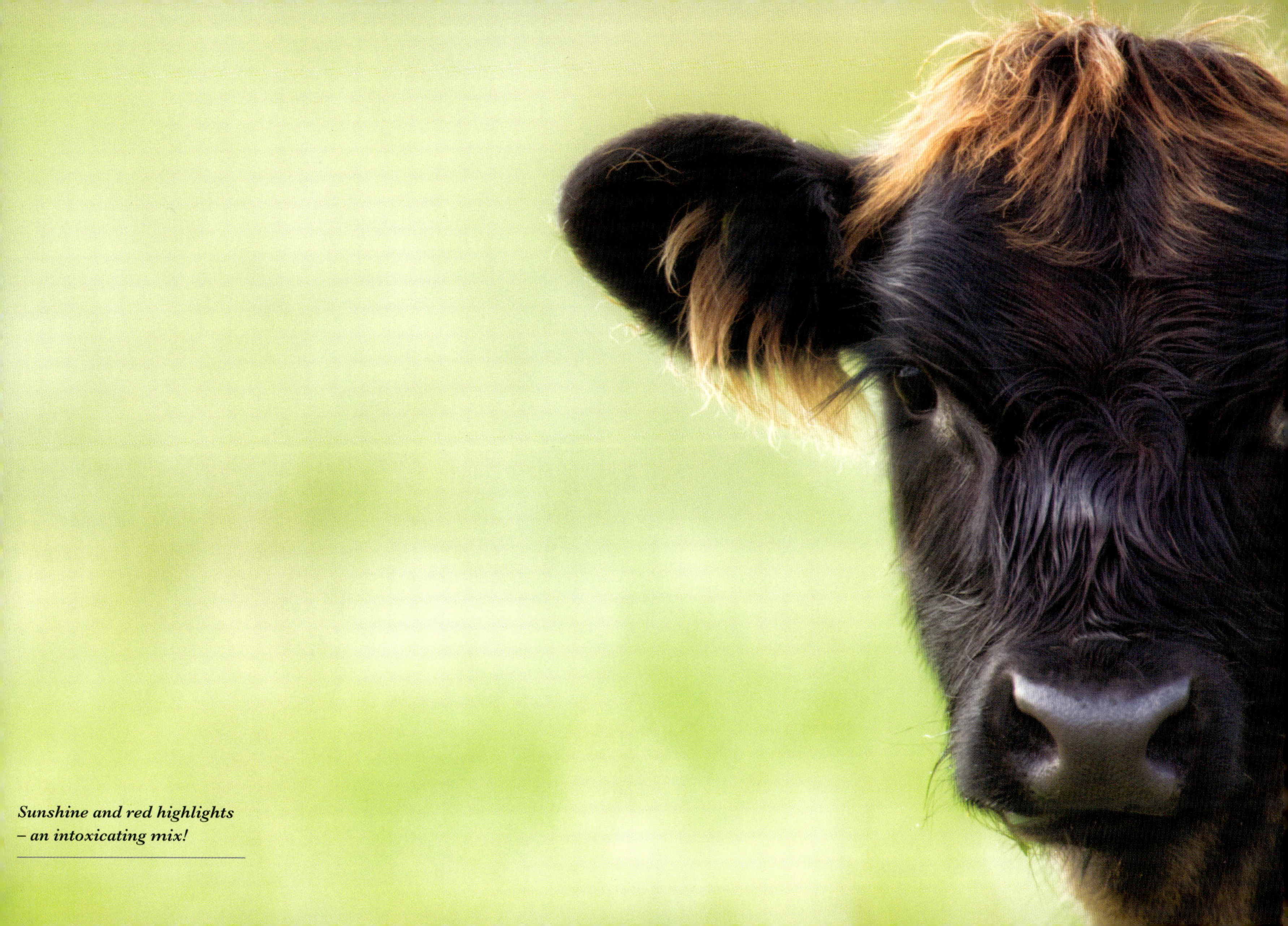

Sunshine and red highlights
– an intoxicating mix!

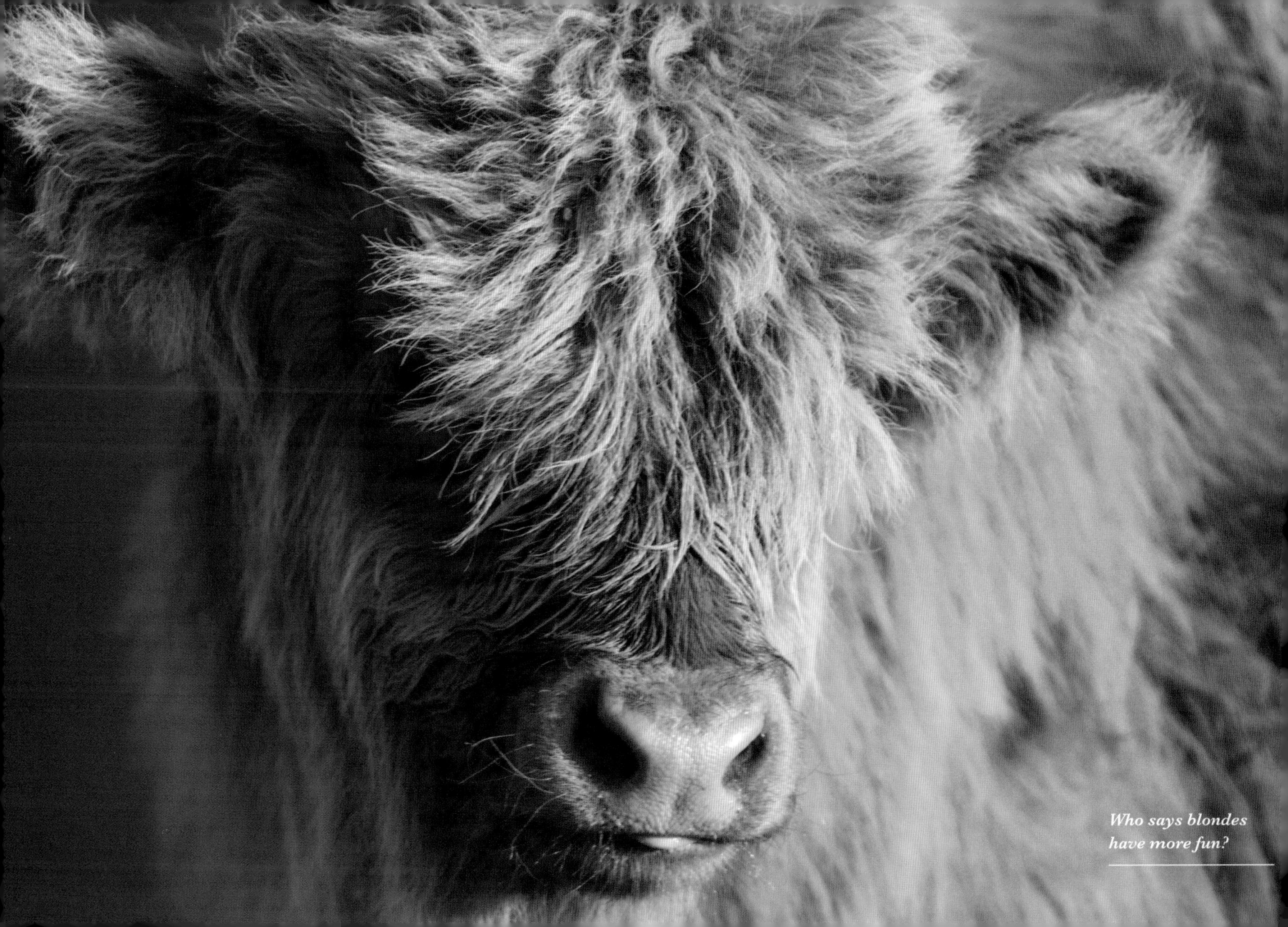
Who says blondes
have more fun?

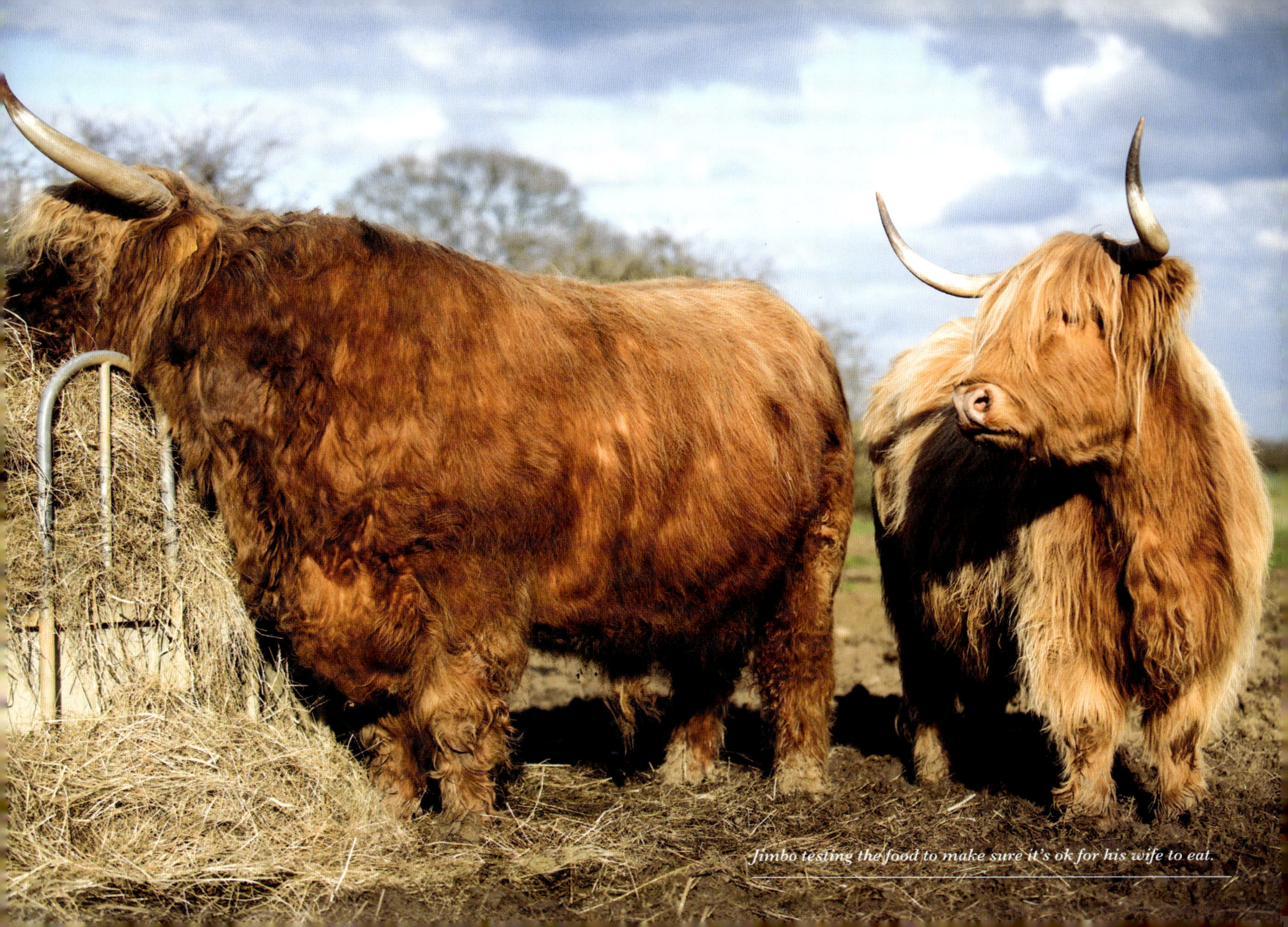

Jimbo testing the food to make sure it's ok for his wife to eat.

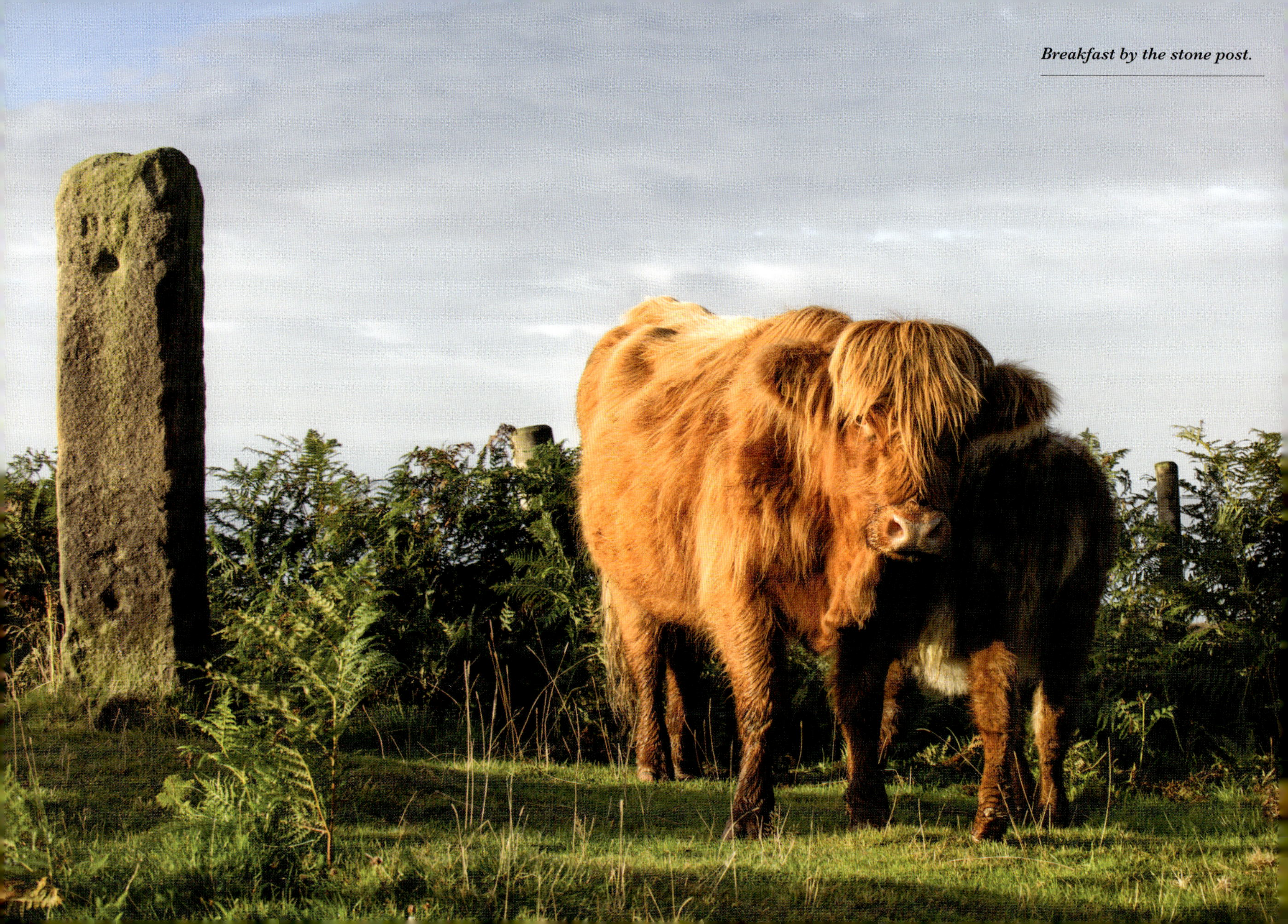

Breakfast by the stone post.

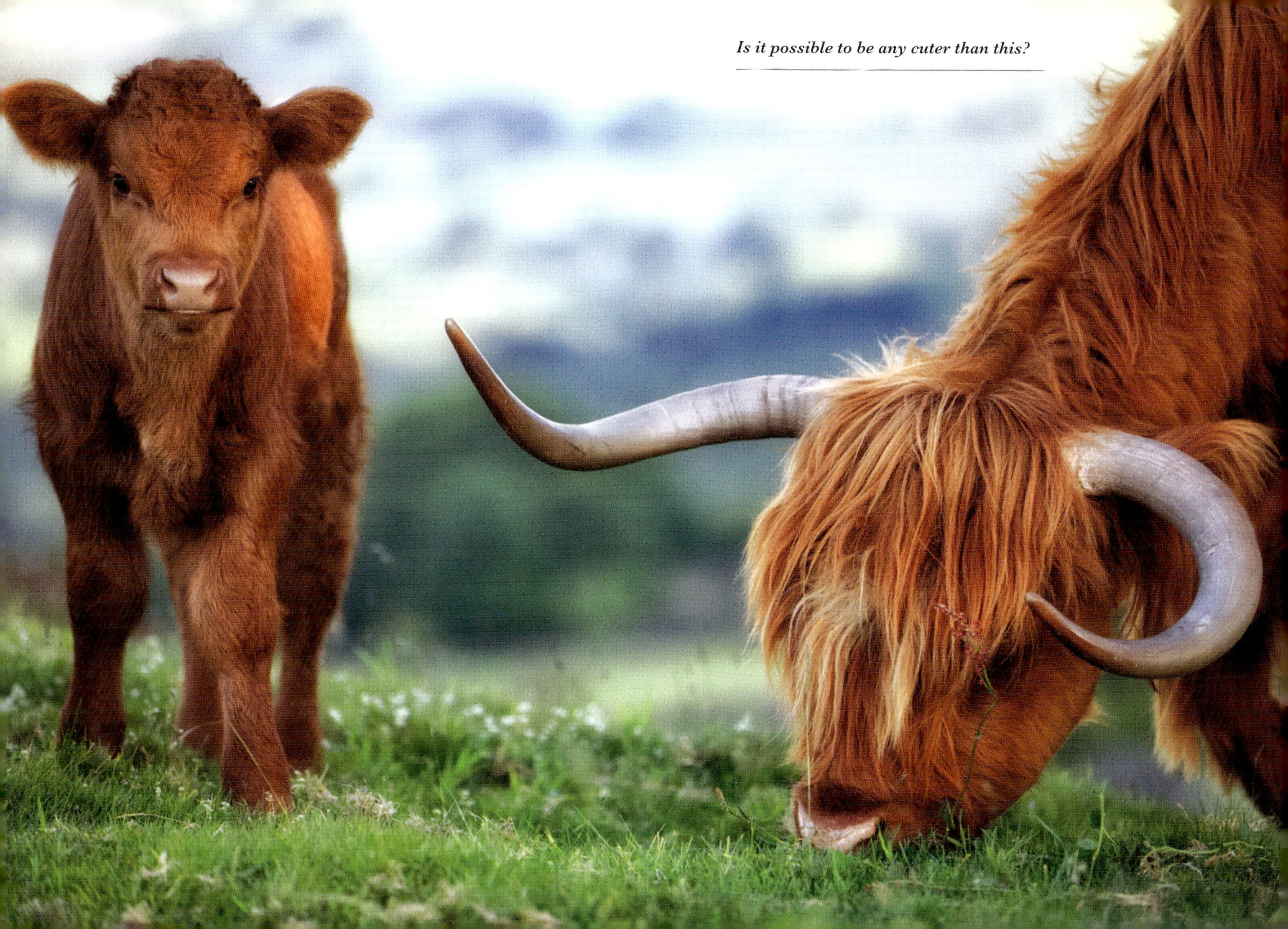
Is it possible to be any cuter than this?

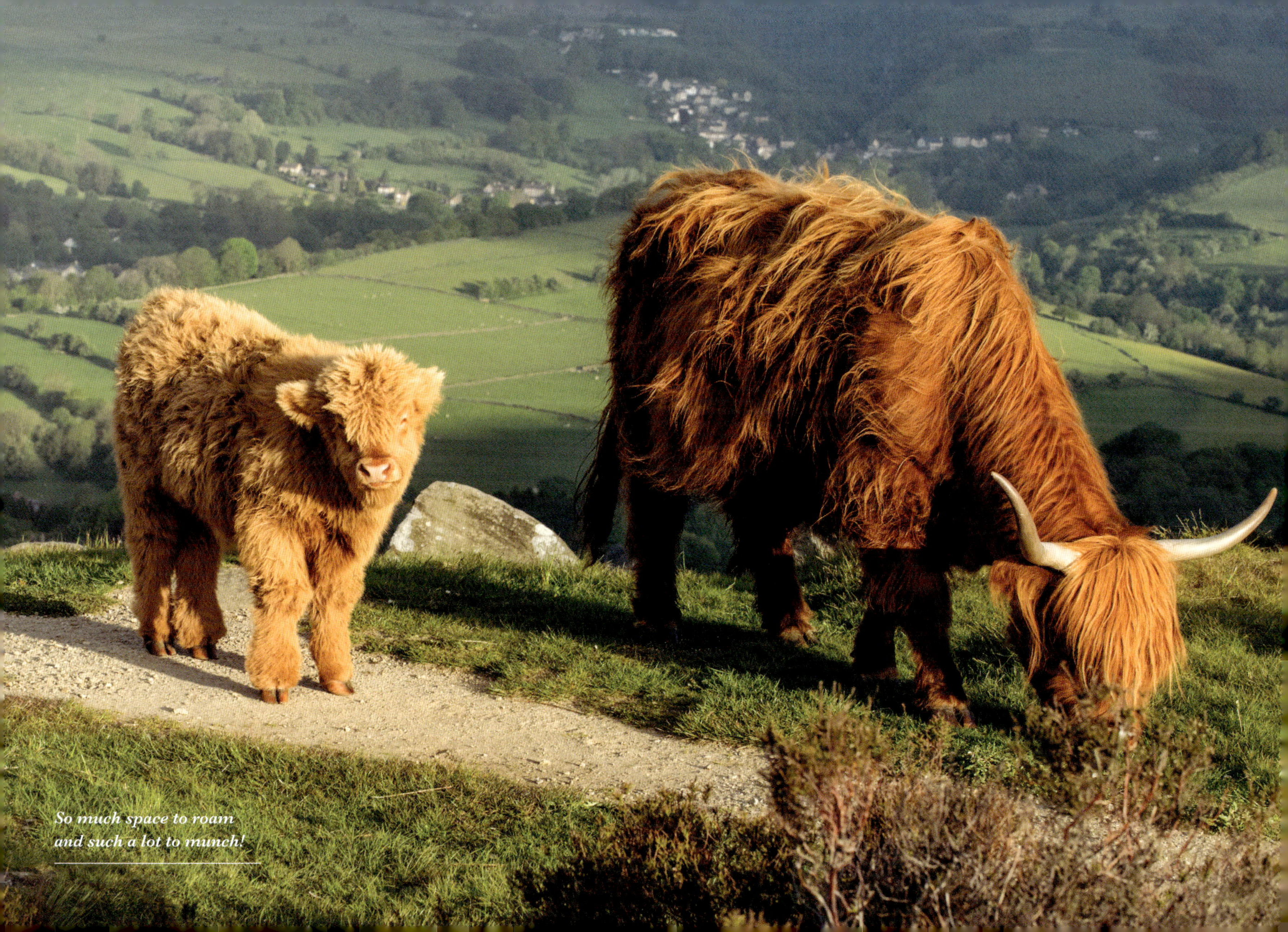
So much space to roam
and such a lot to munch!

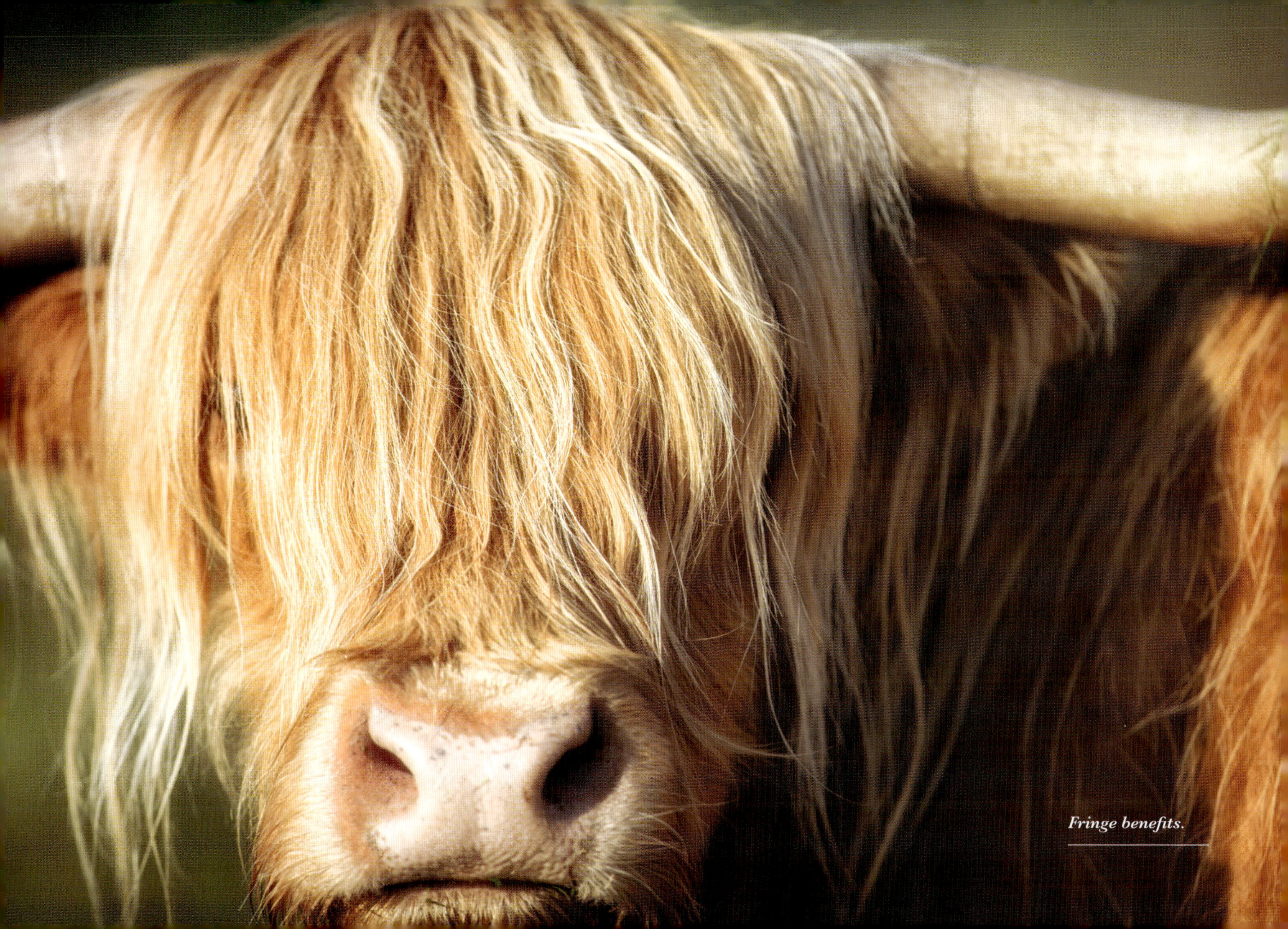

Fringe benefits.

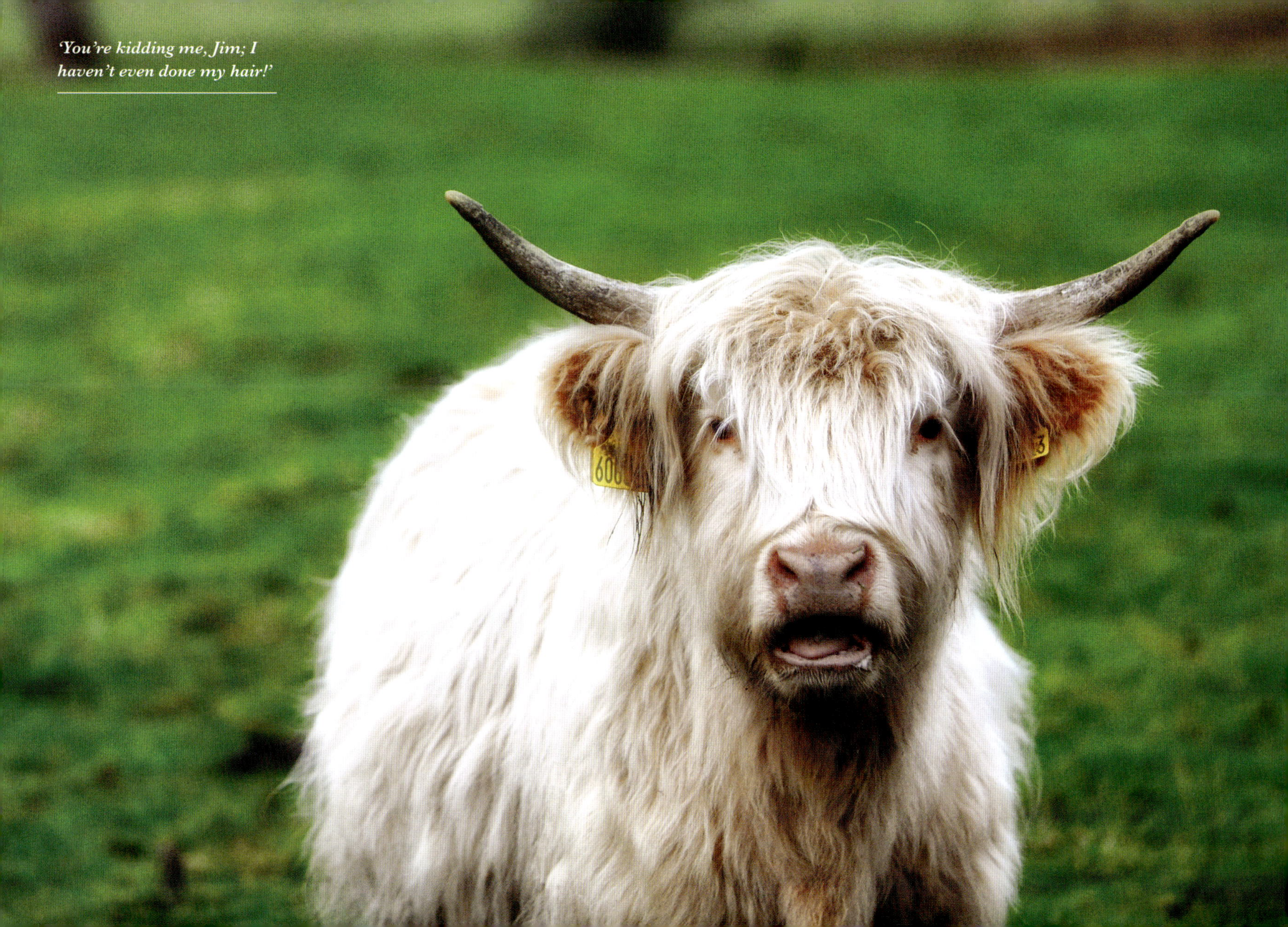

'You're kidding me, Jim; I haven't even done my hair!'

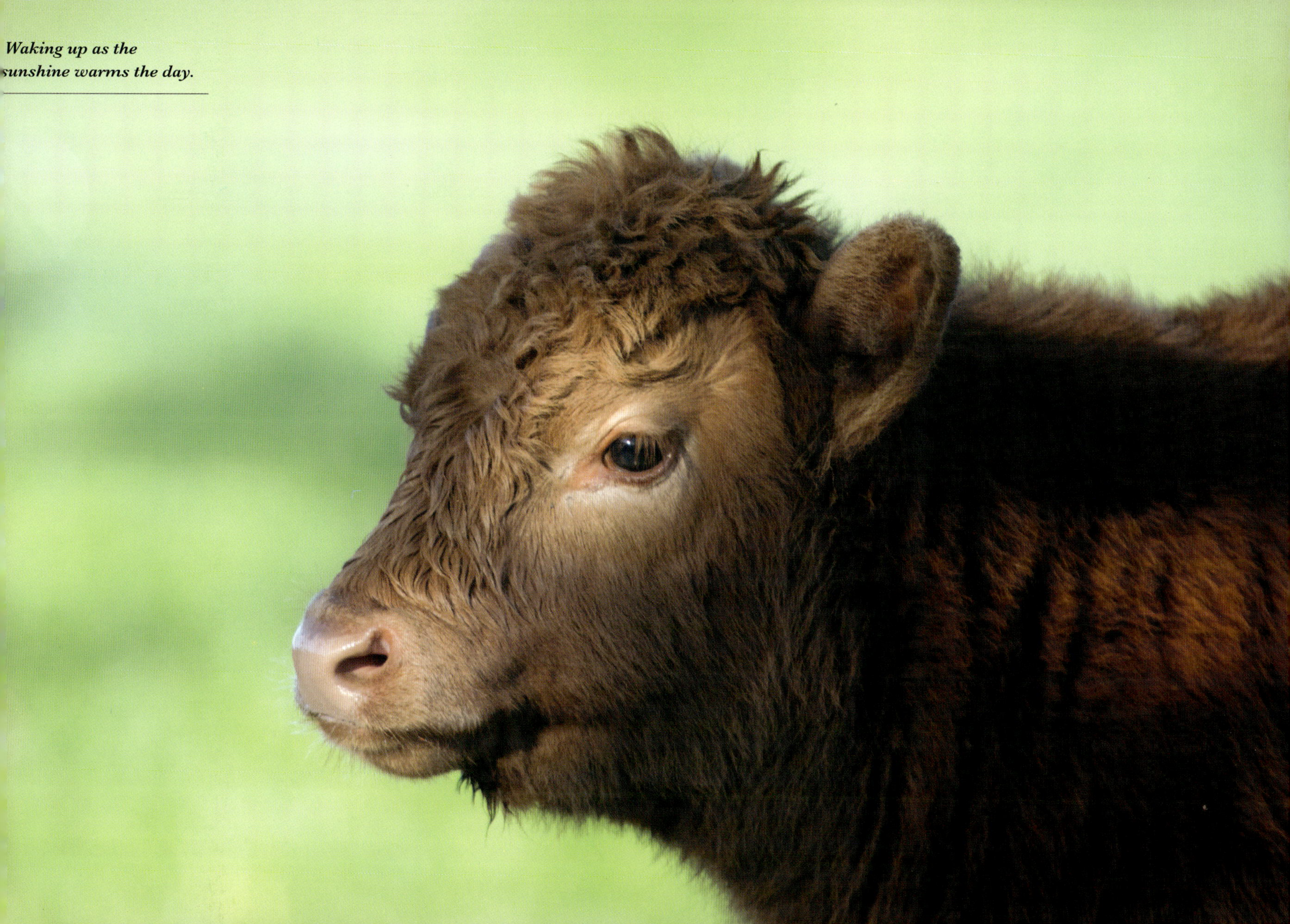Waking up as the
sunshine warms the day.

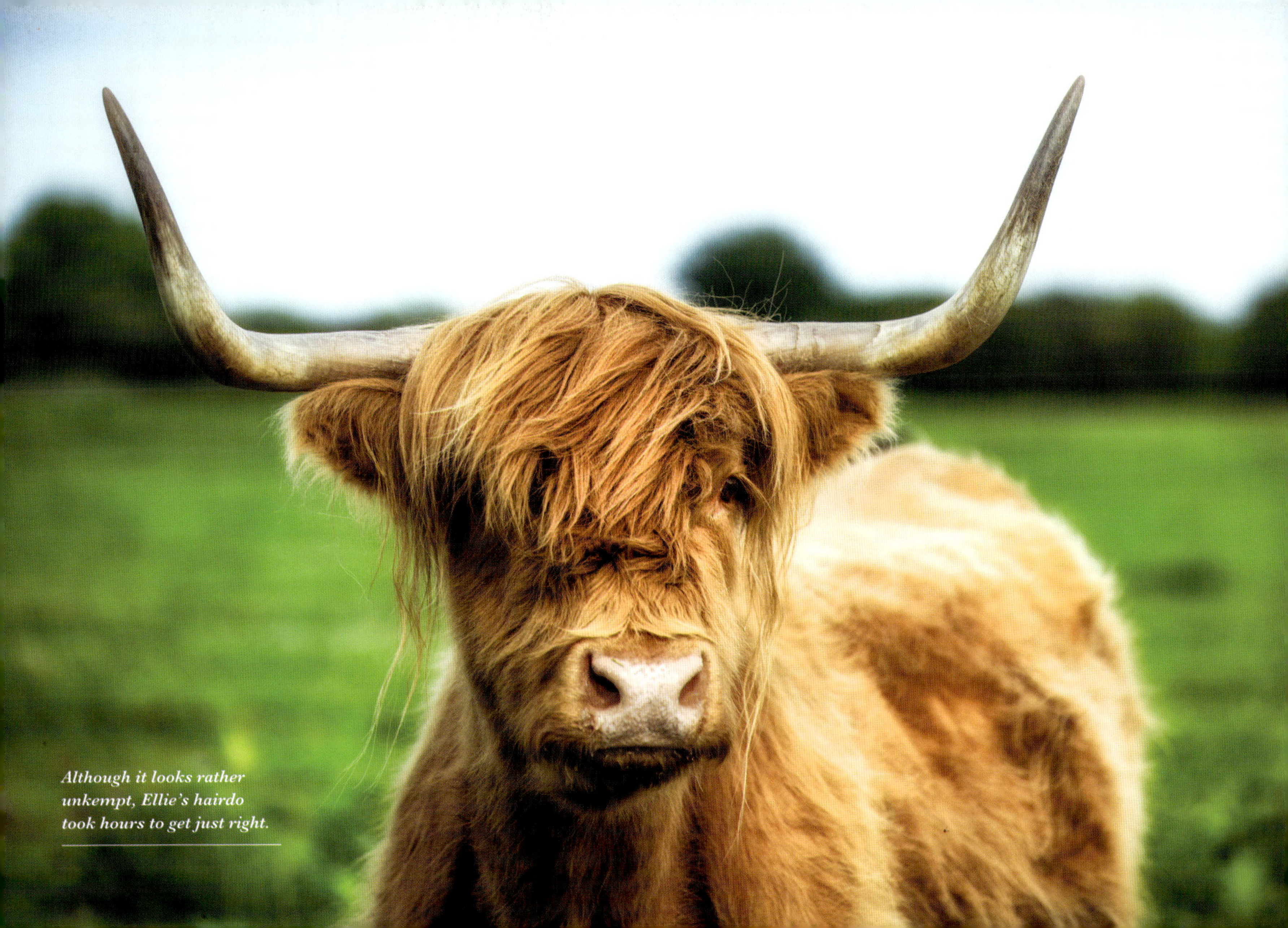

Although it looks rather
unkempt, Ellie's hairdo
took hours to get just right.

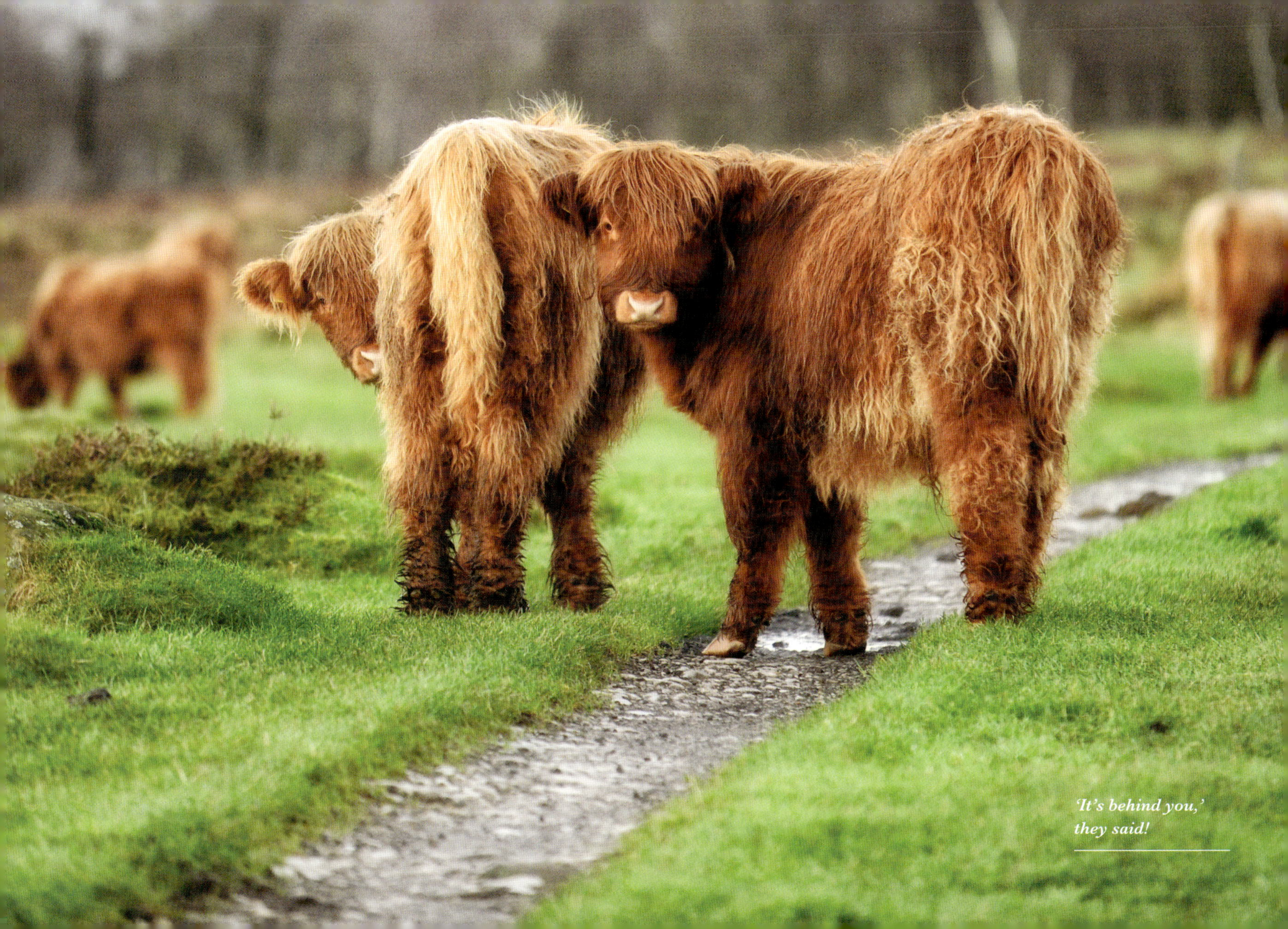

'It's behind you,'
they said!

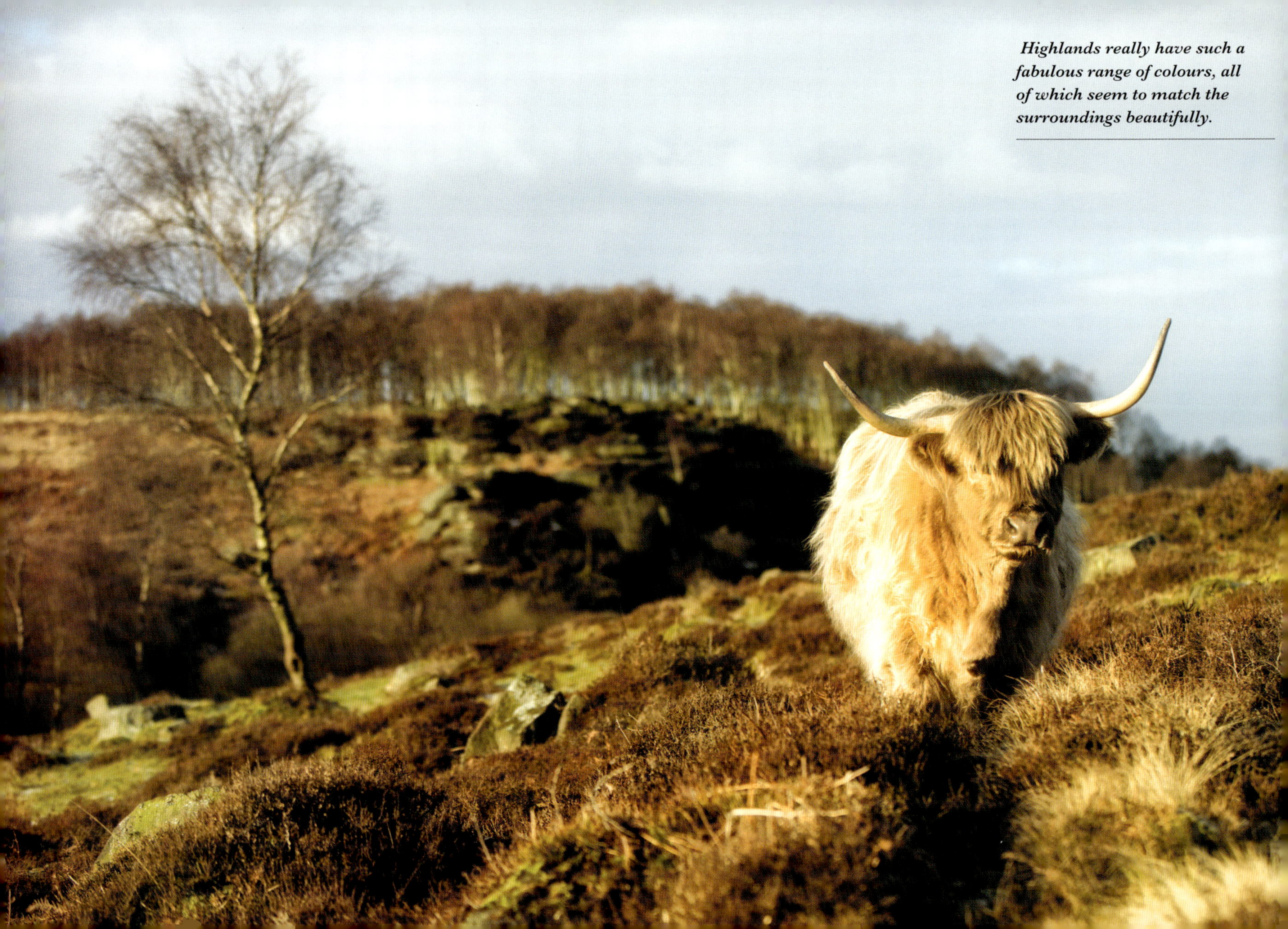
Highlands really have such a
fabulous range of colours, all
of which seem to match the
surroundings beautifully.

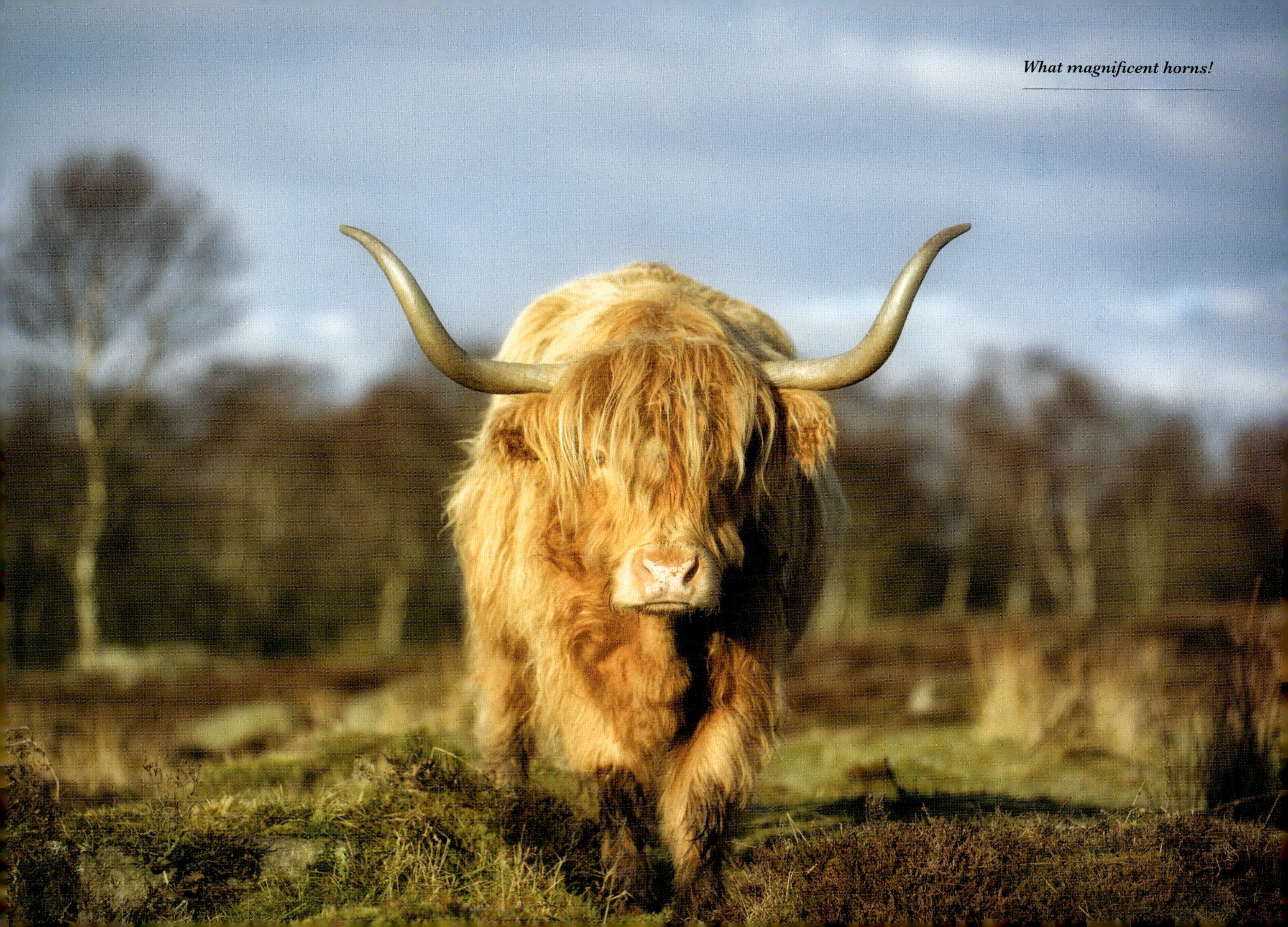

What magnificent horns!

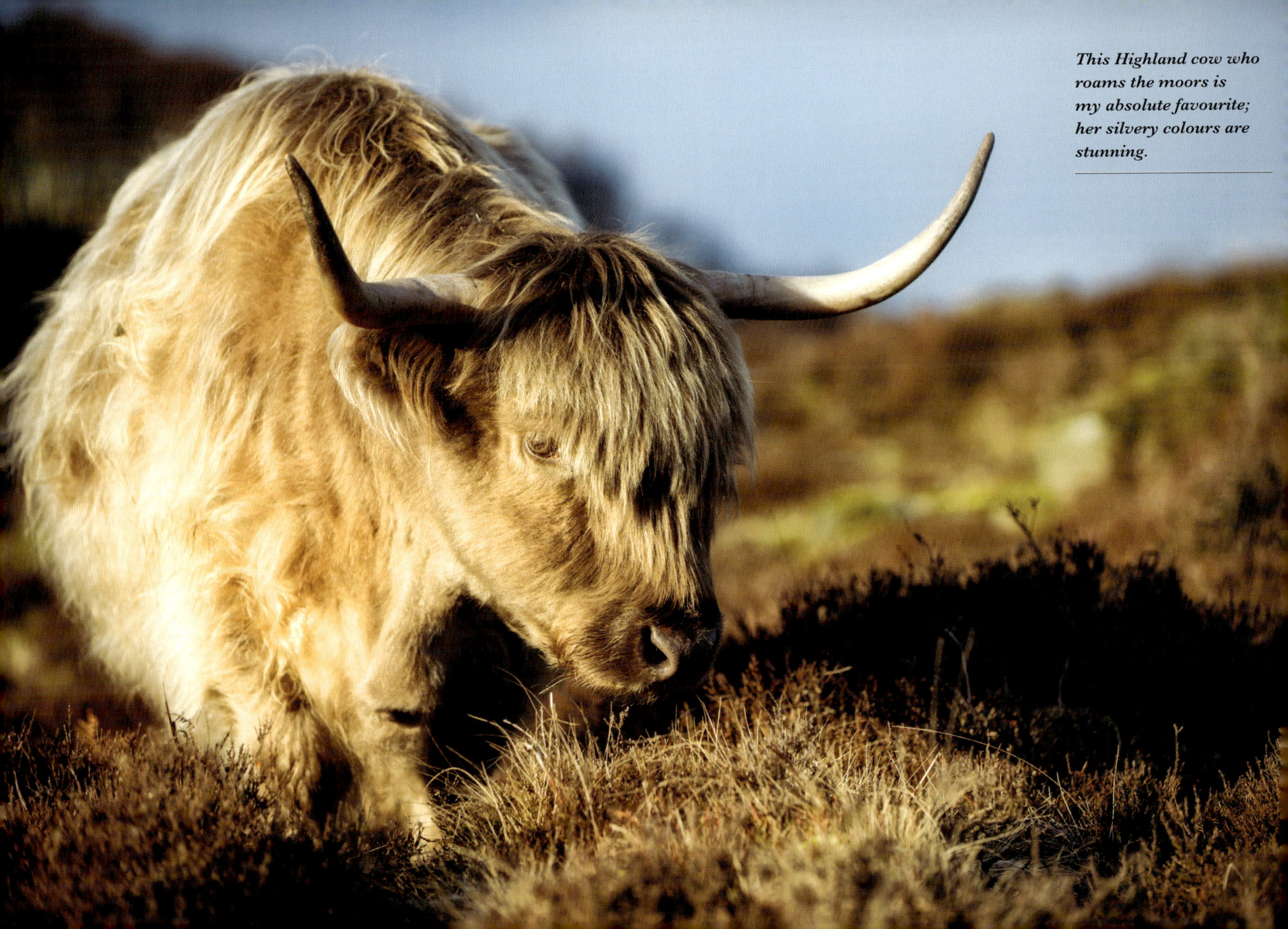

This Highland cow who roams the moors is my absolute favourite; her silvery colours are stunning.

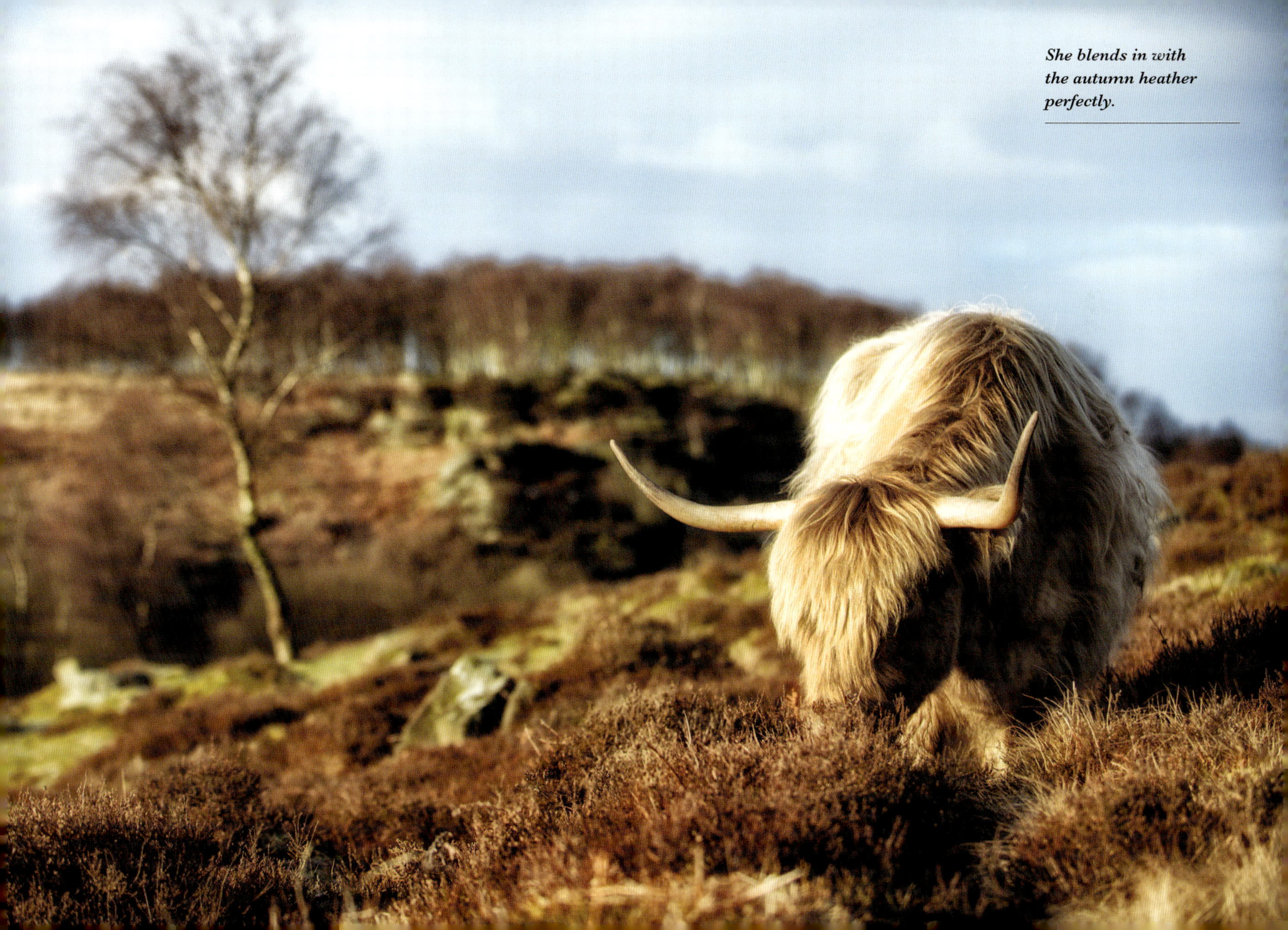
She blends in with
the autumn heather
perfectly.

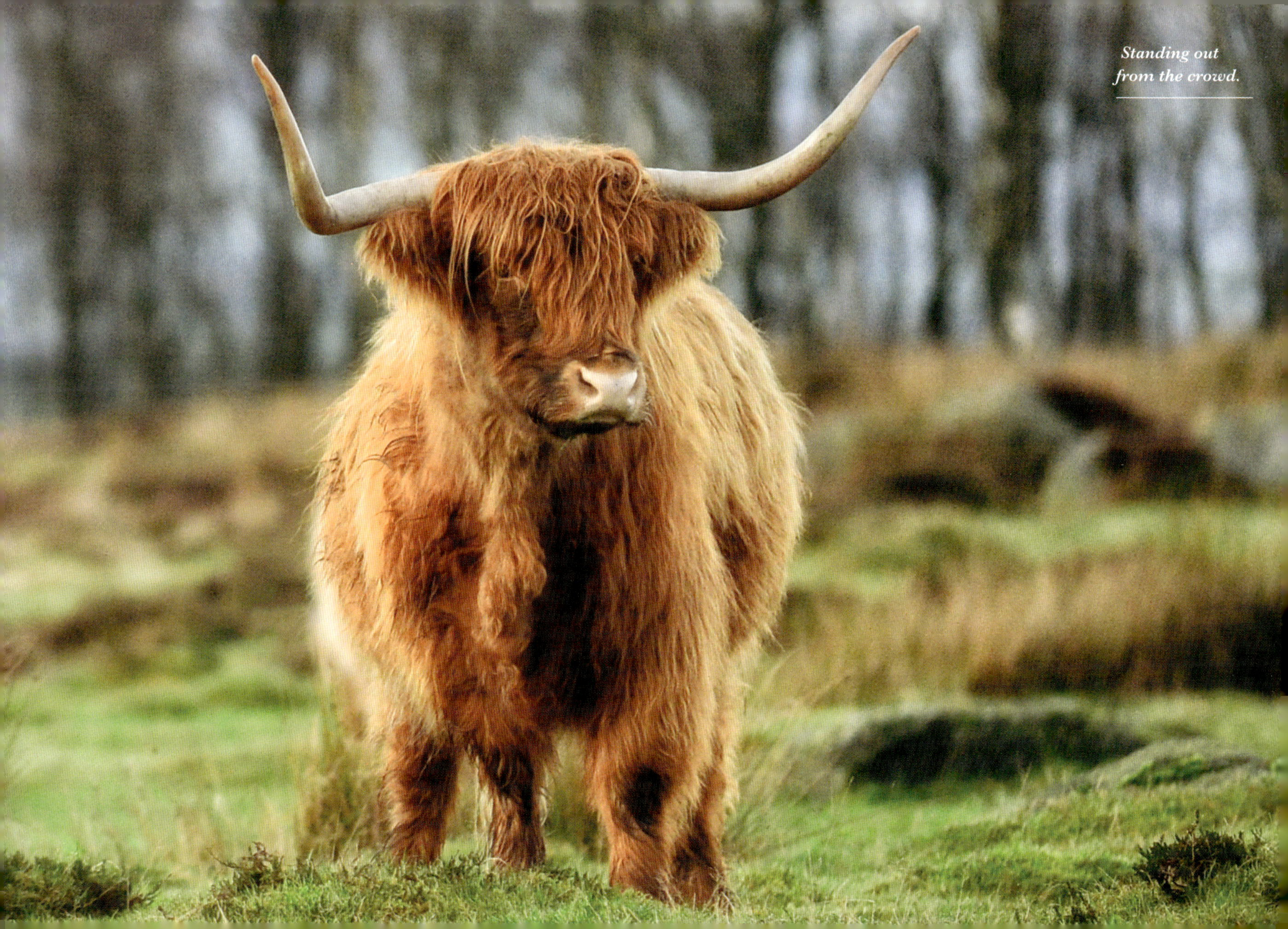

Standing out
from the crowd.

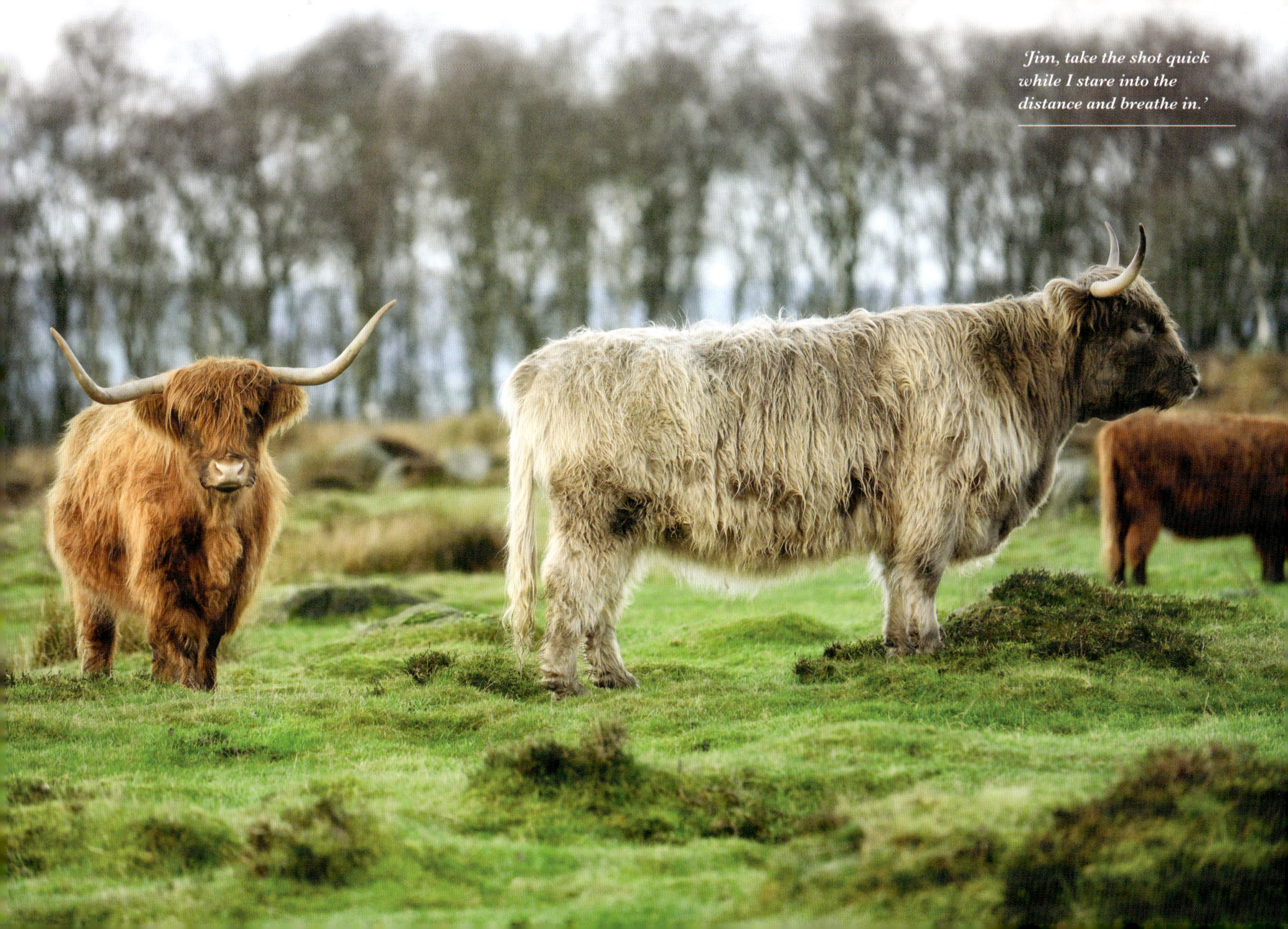
'Jim, take the shot quick
while I stare into the
distance and breathe in.'

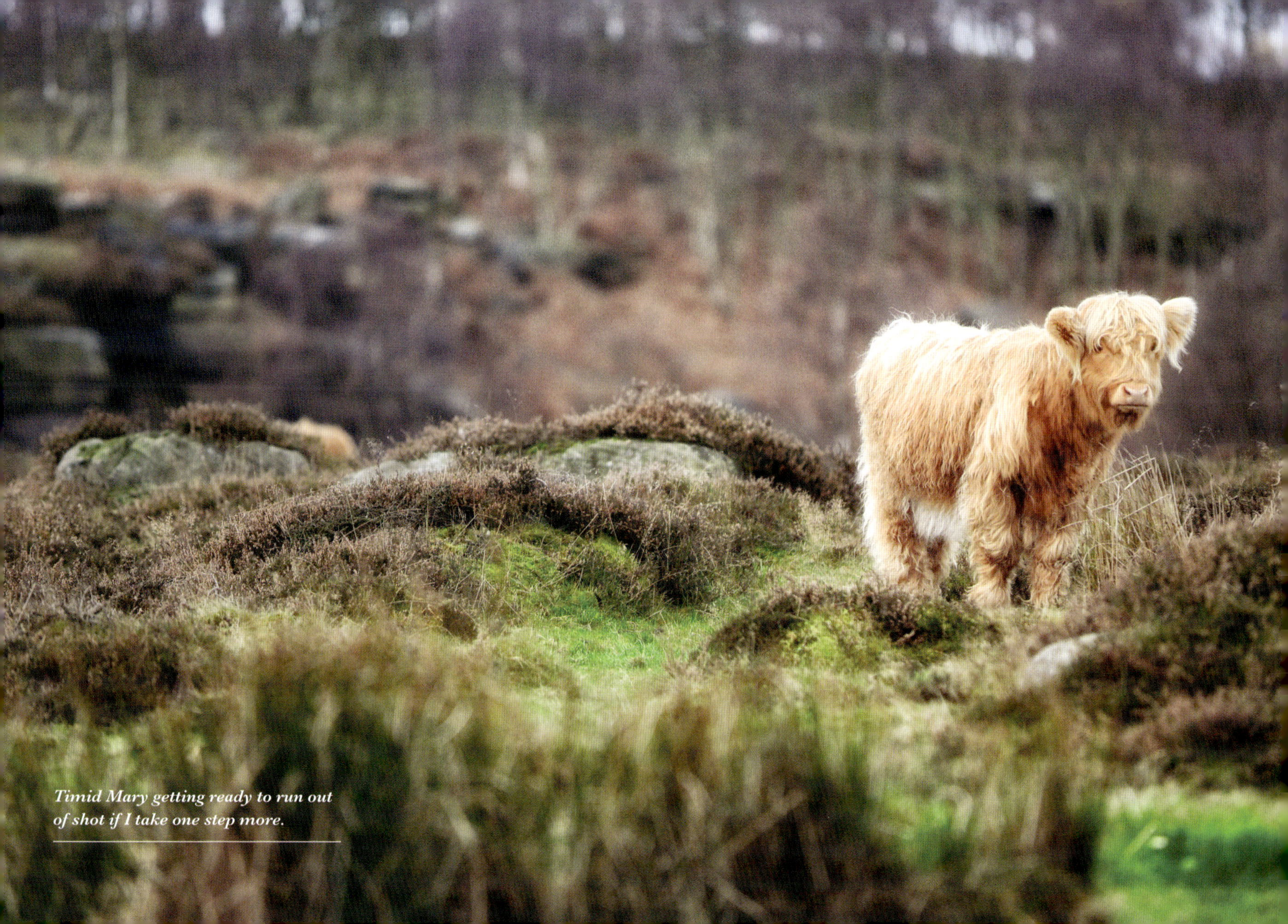

Timid Mary getting ready to run out of shot if I take one step more.

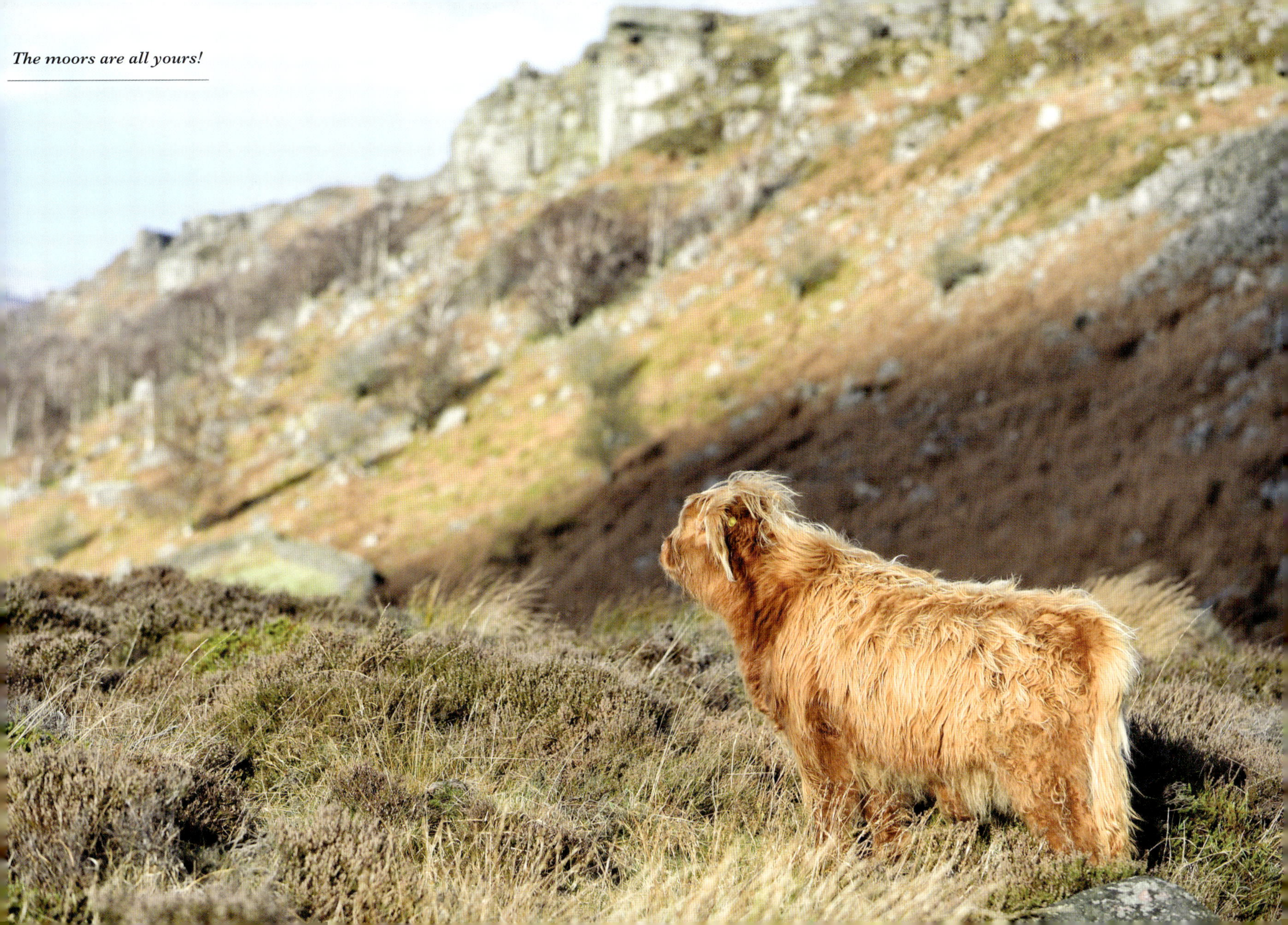
The moors are all yours!

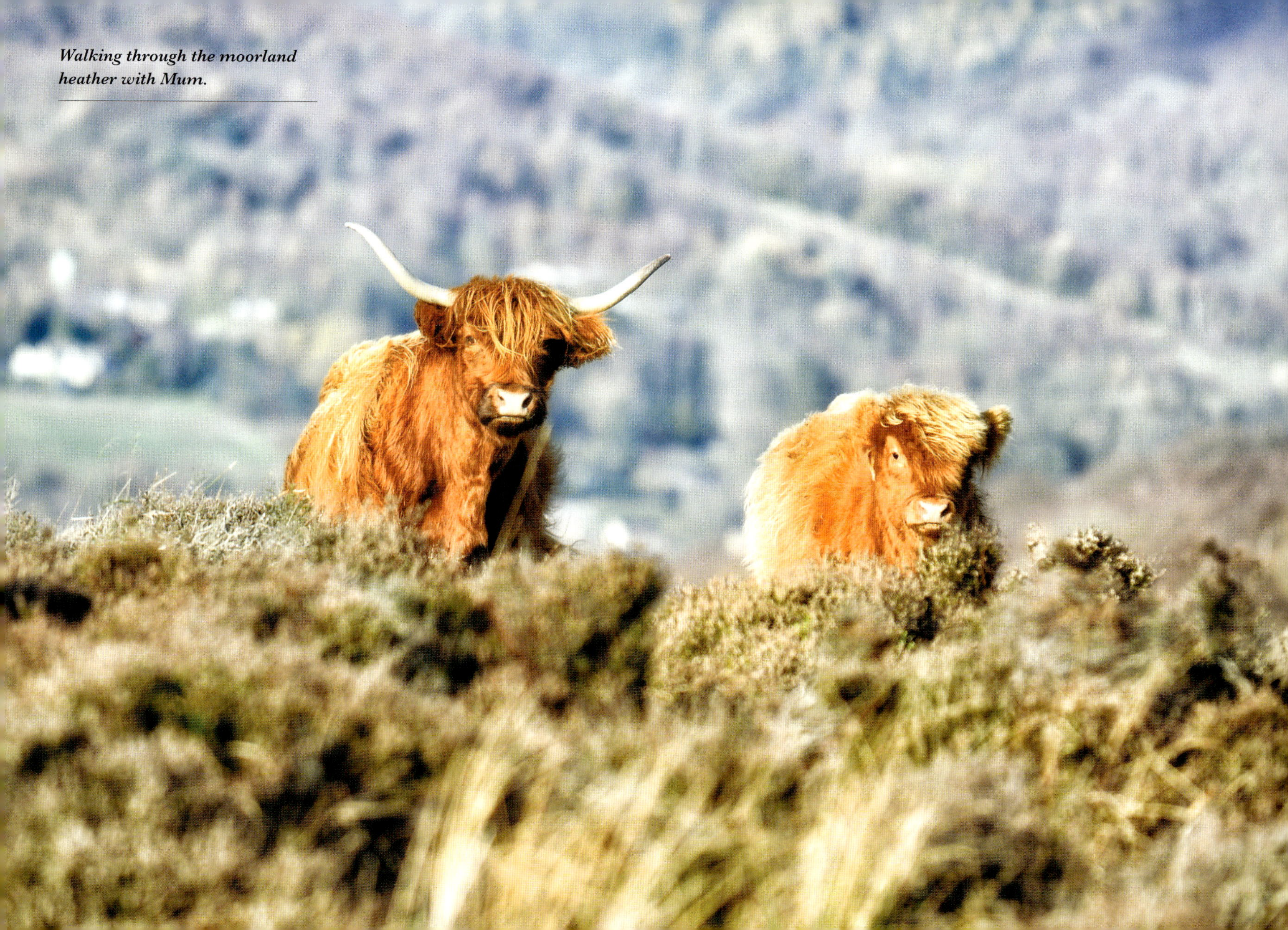

Walking through the moorland heather with Mum.

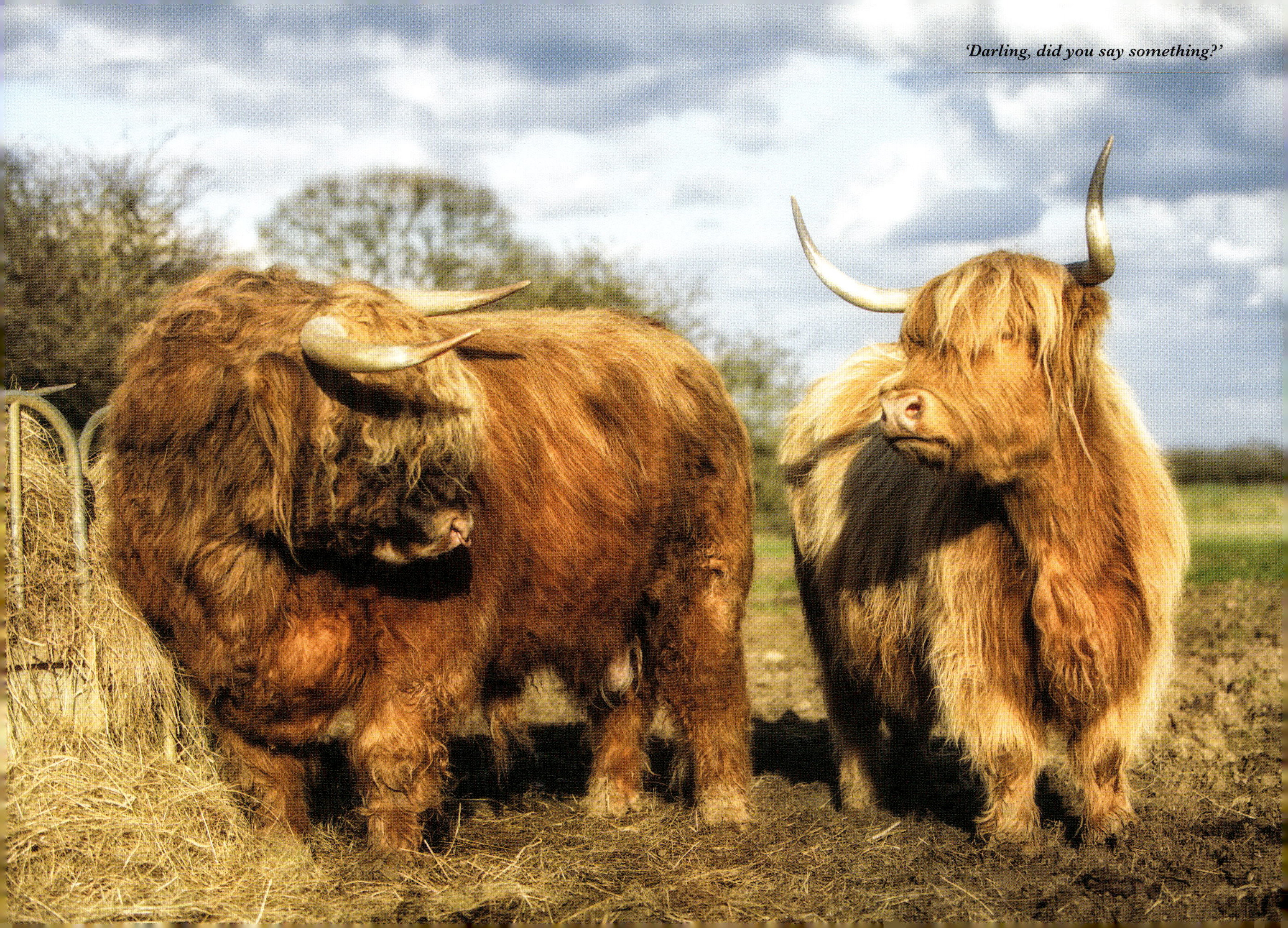

'Darling, did you say something?'

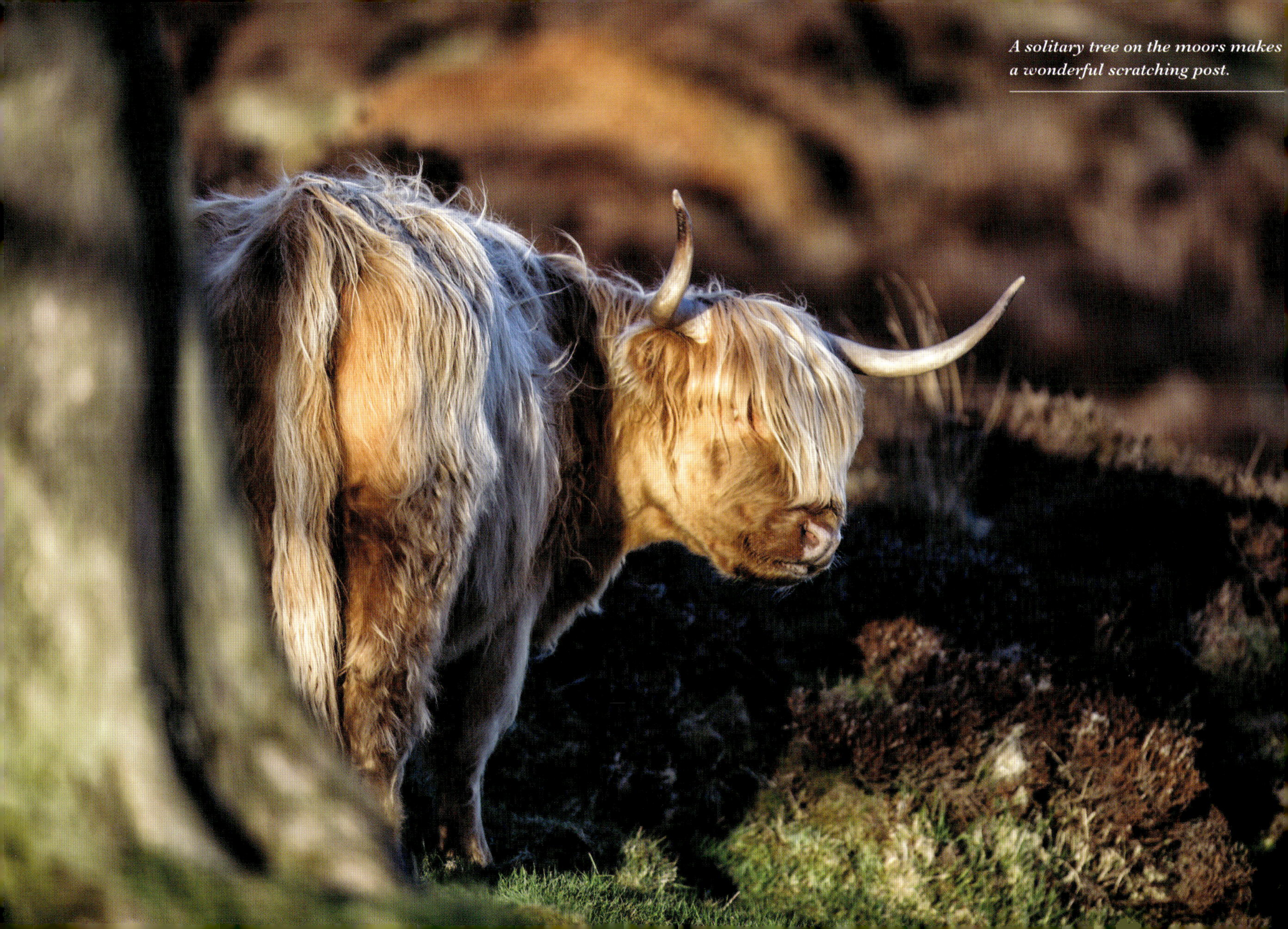

A solitary tree on the moors makes
a wonderful scratching post.

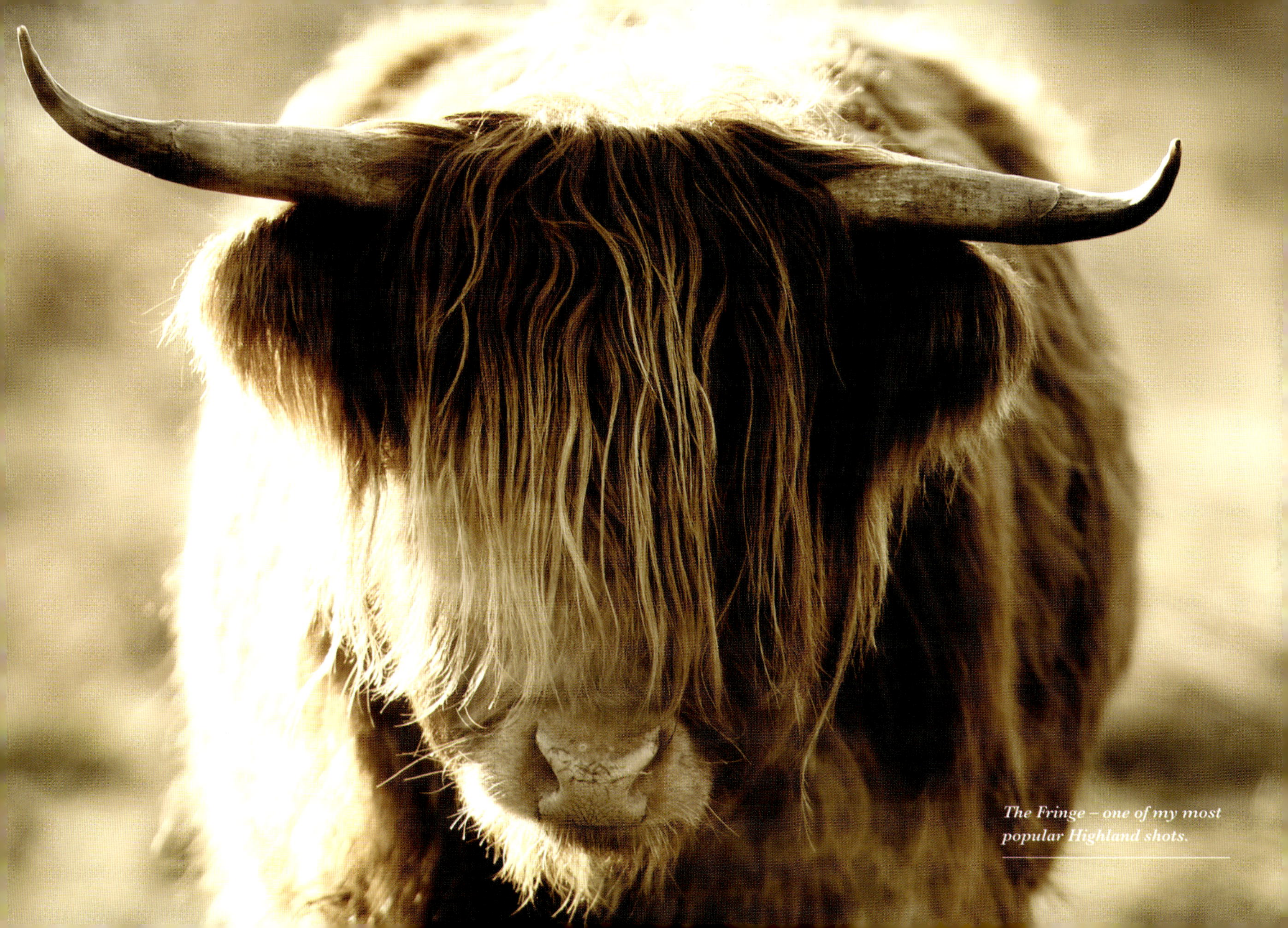

The Fringe – one of my most
popular Highland shots.

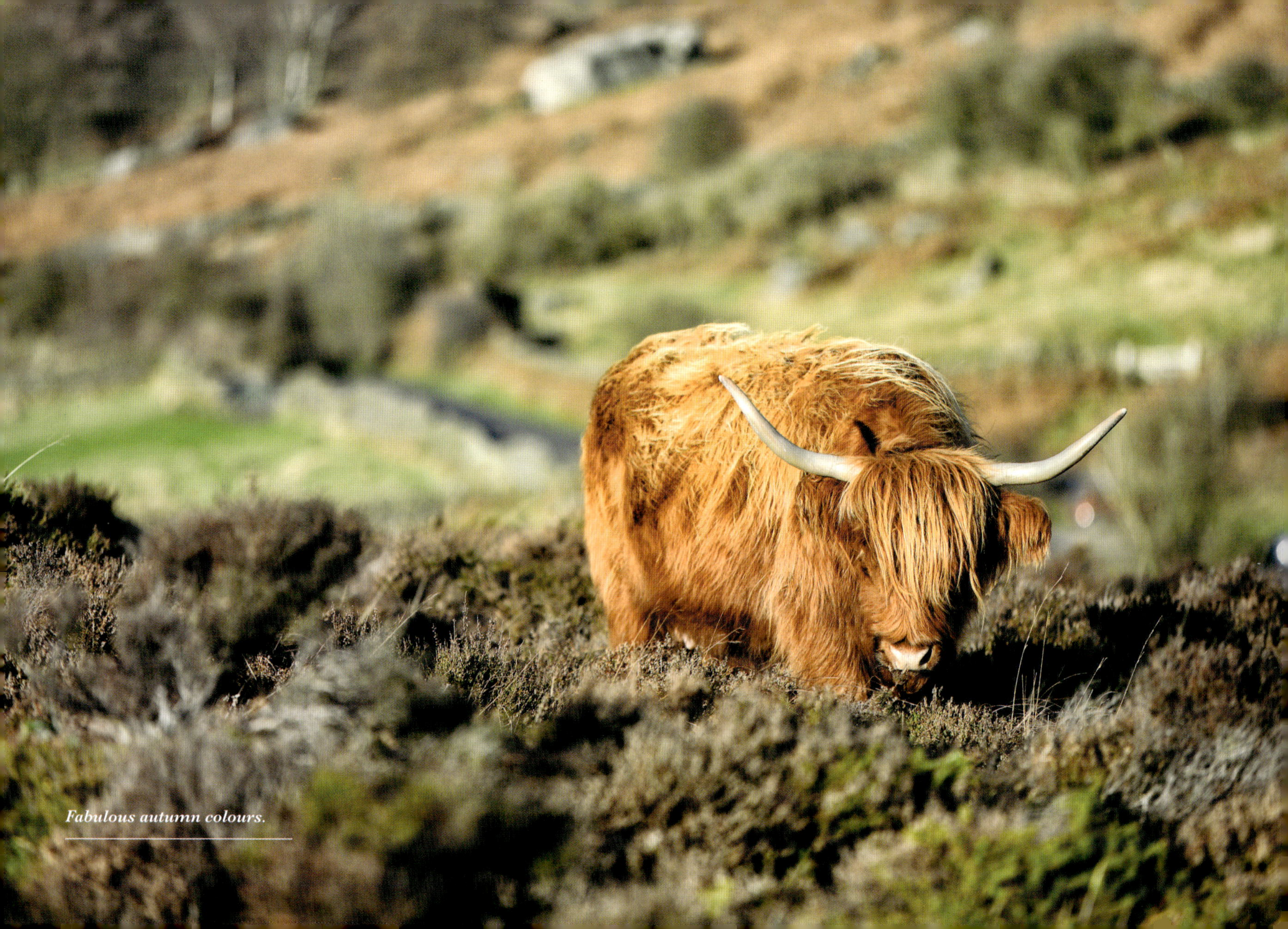

Fabulous autumn colours.

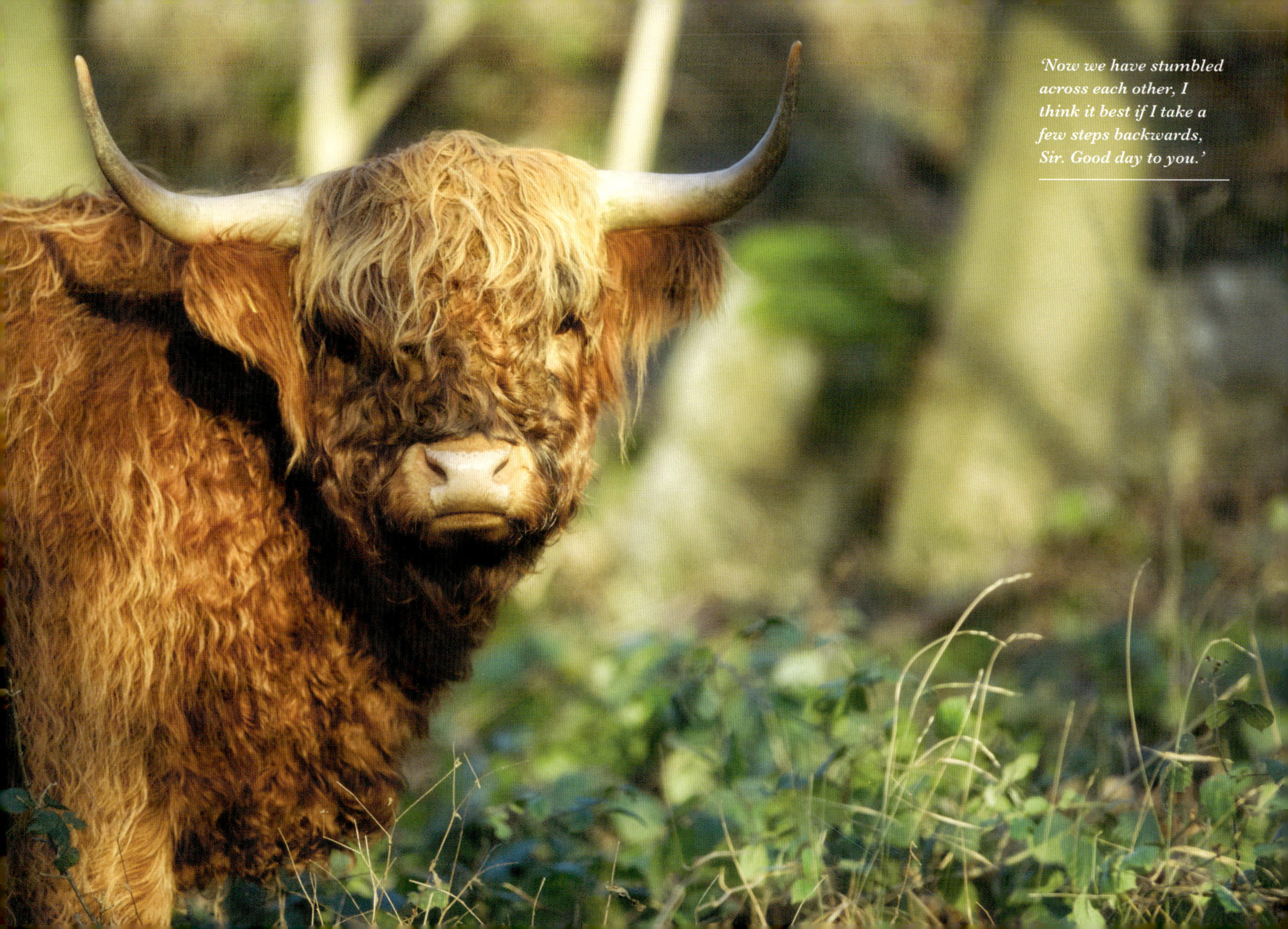
'Now we have stumbled
across each other, I
think it best if I take a
few steps backwards,
Sir. Good day to you.'

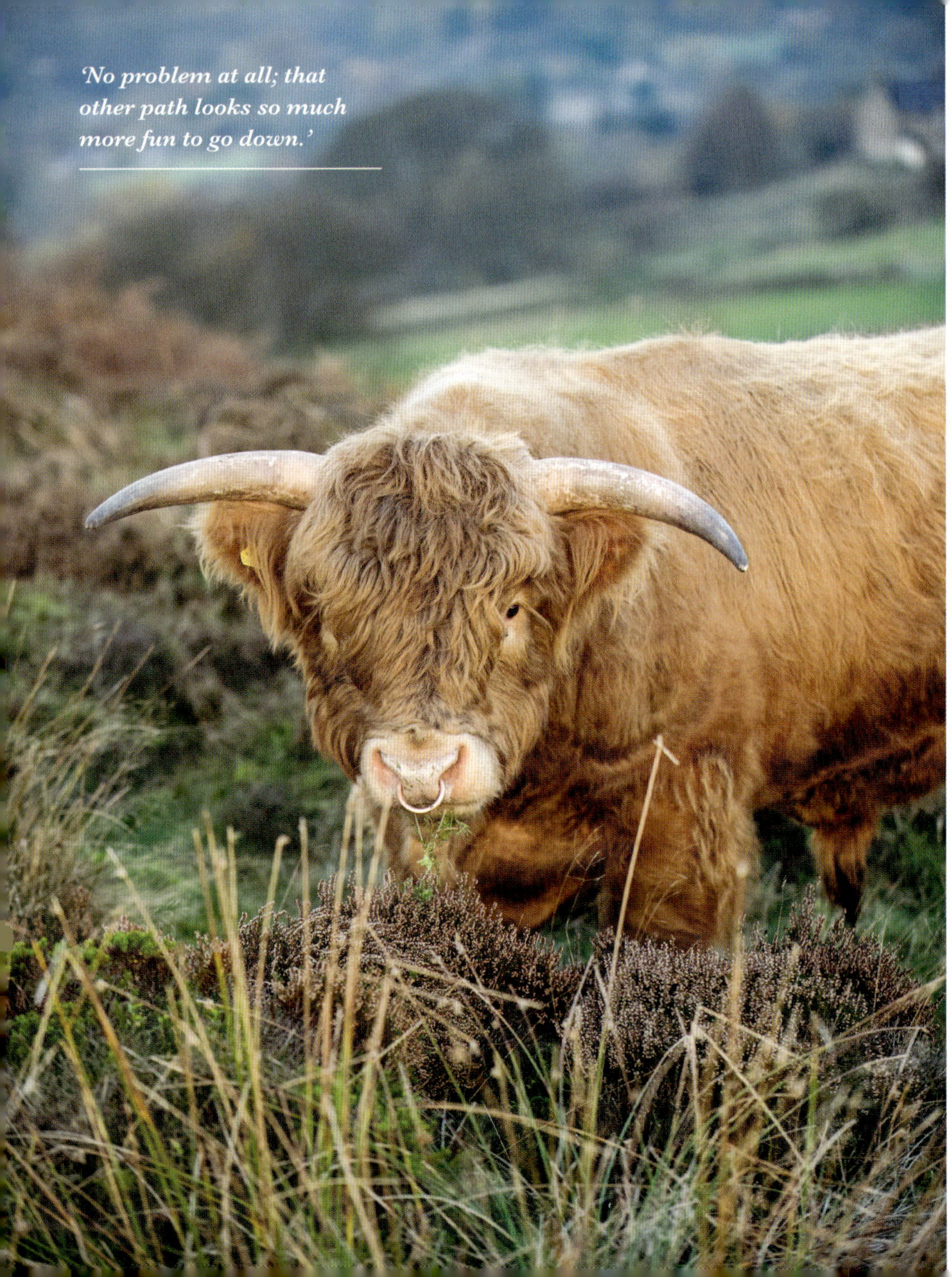

'No problem at all; that other path looks so much more fun to go down.'

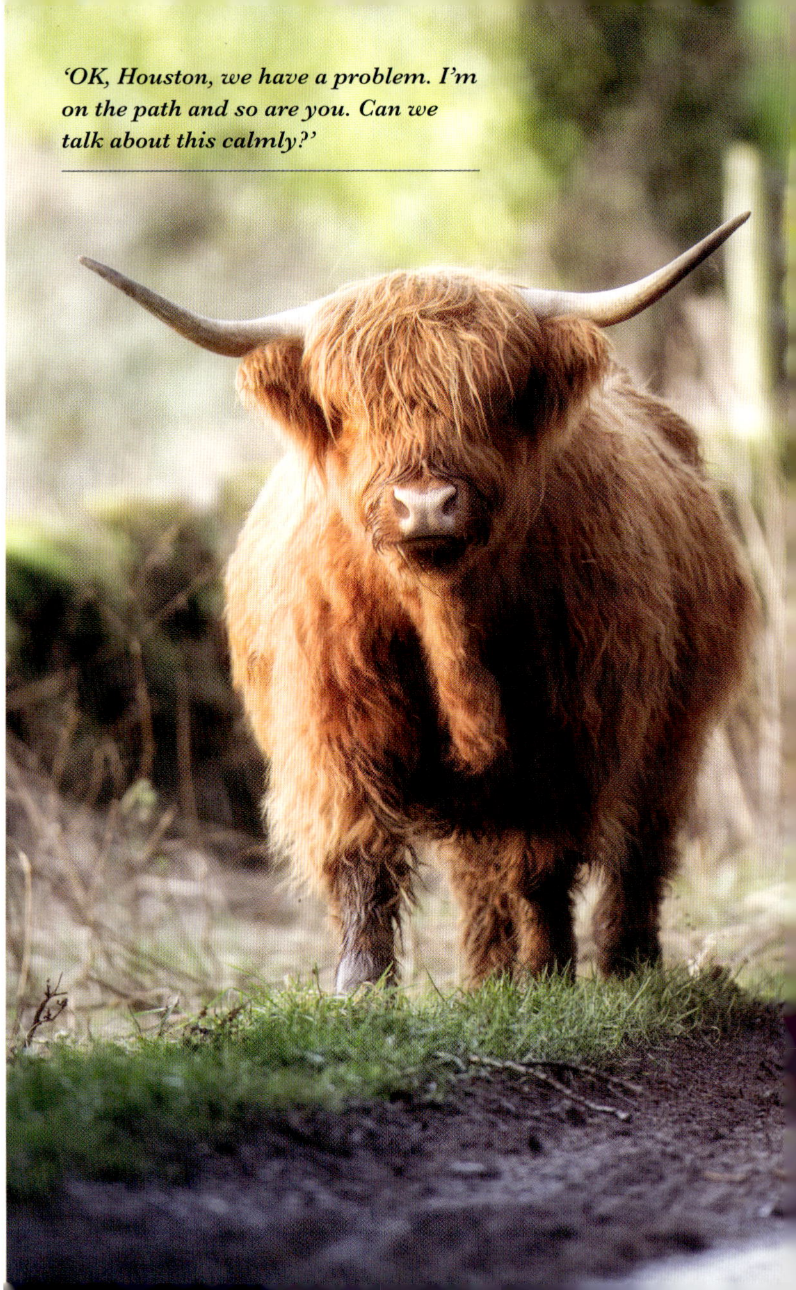

'OK, Houston, we have a problem. I'm on the path and so are you. Can we talk about this calmly?'

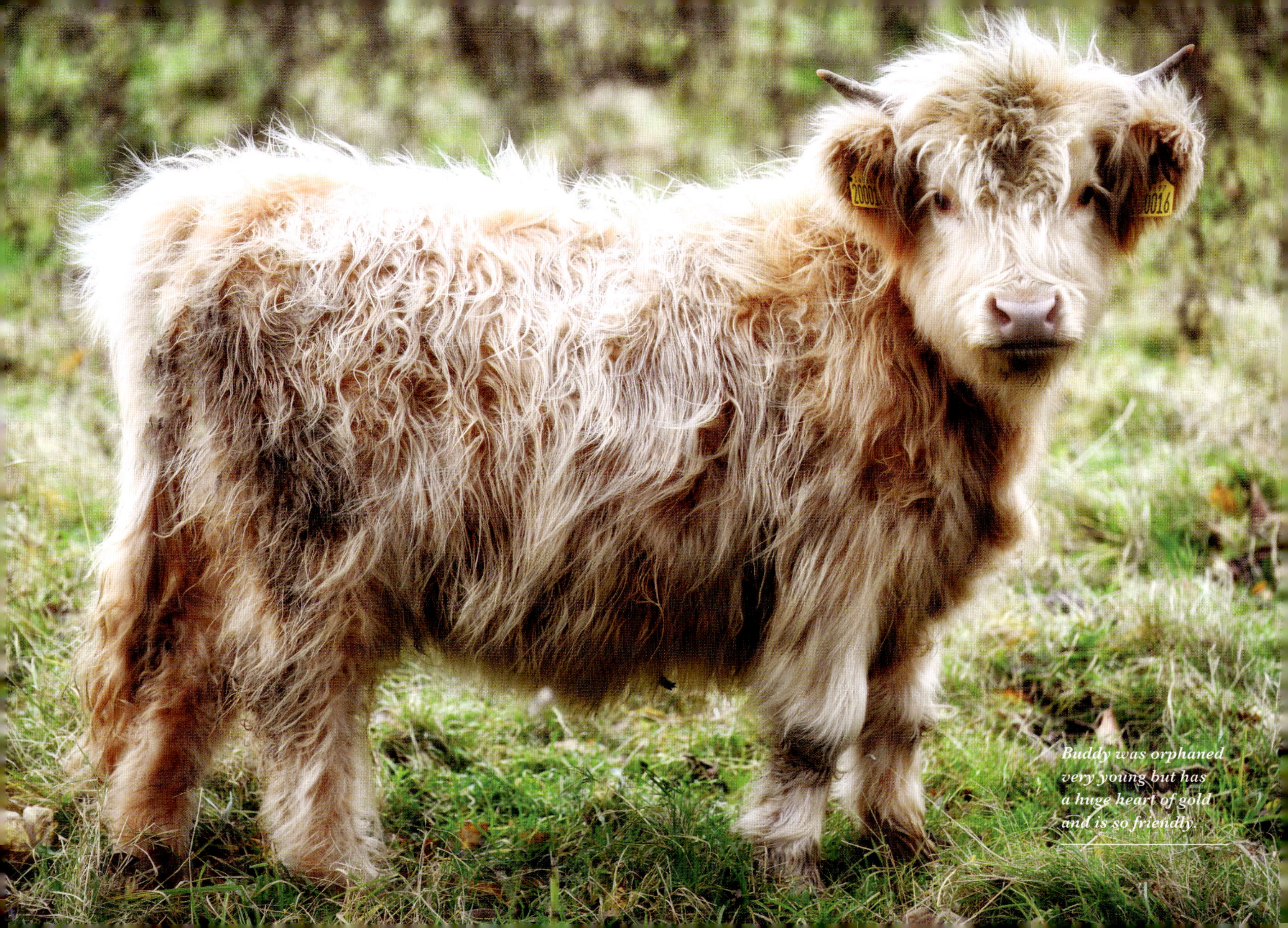

Buddy was orphaned
very young but has
a huge heart of gold
and is so friendly.

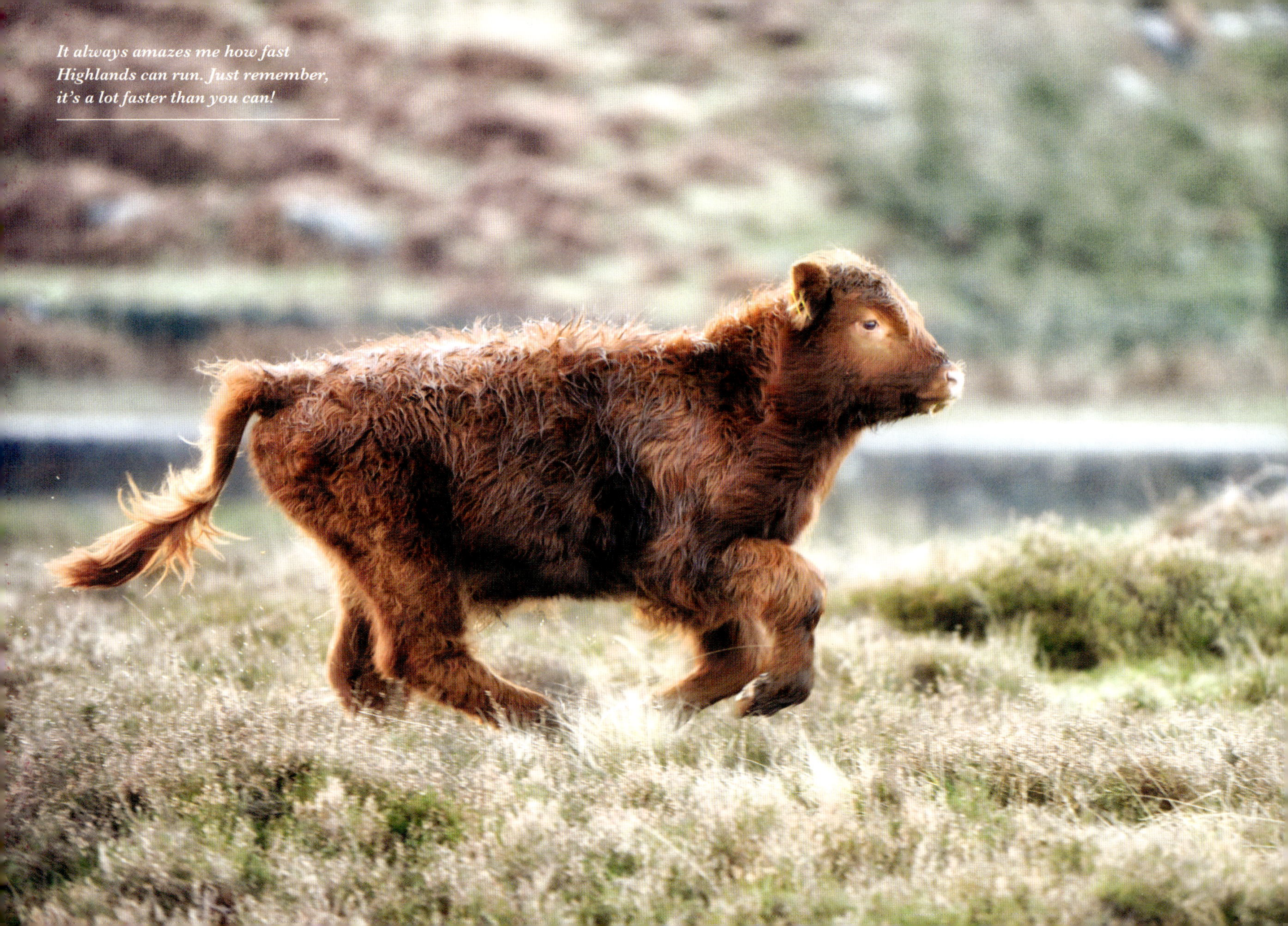
It always amazes me how fast
Highlands can run. Just remember,
it's a lot faster than you can!

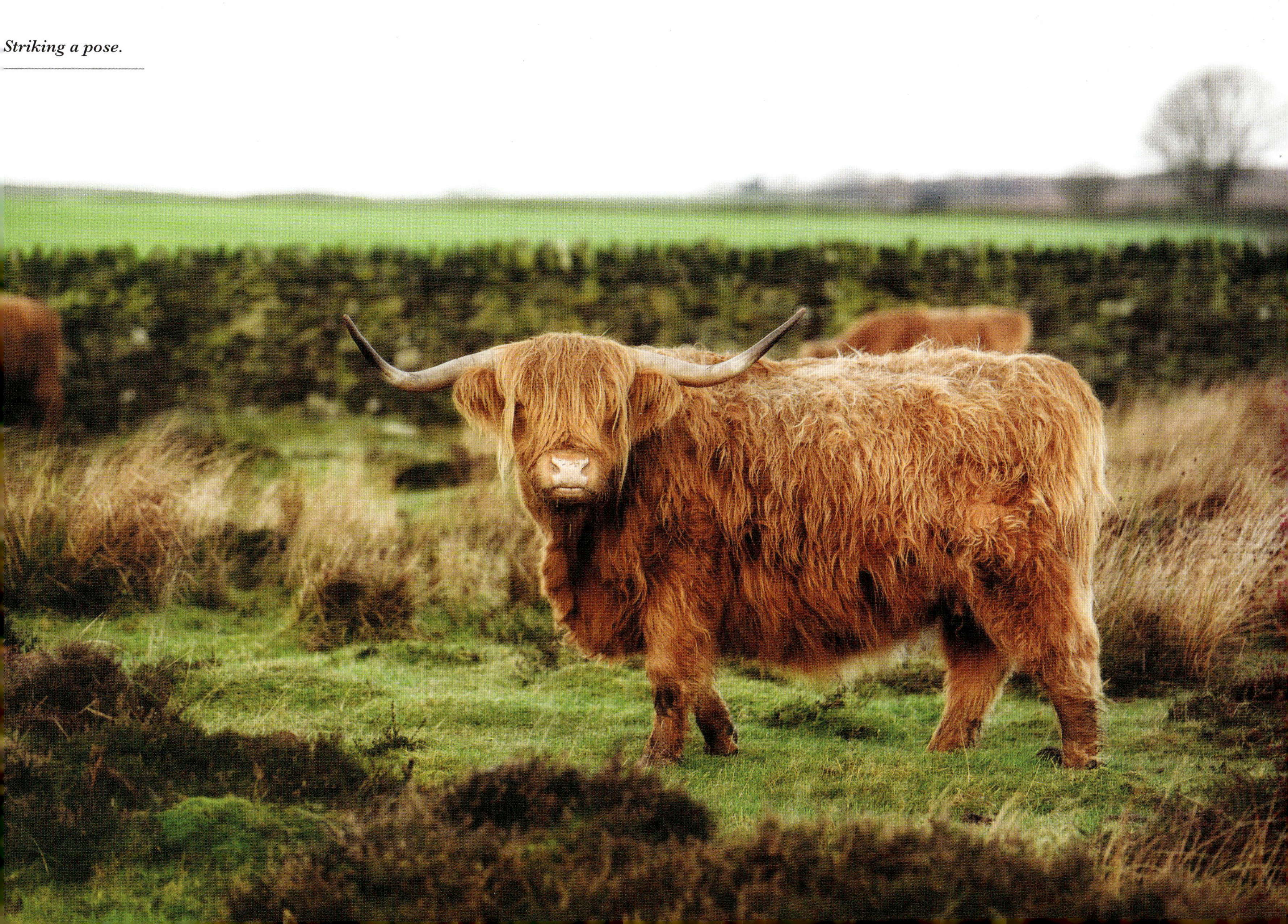

Striking a pose.

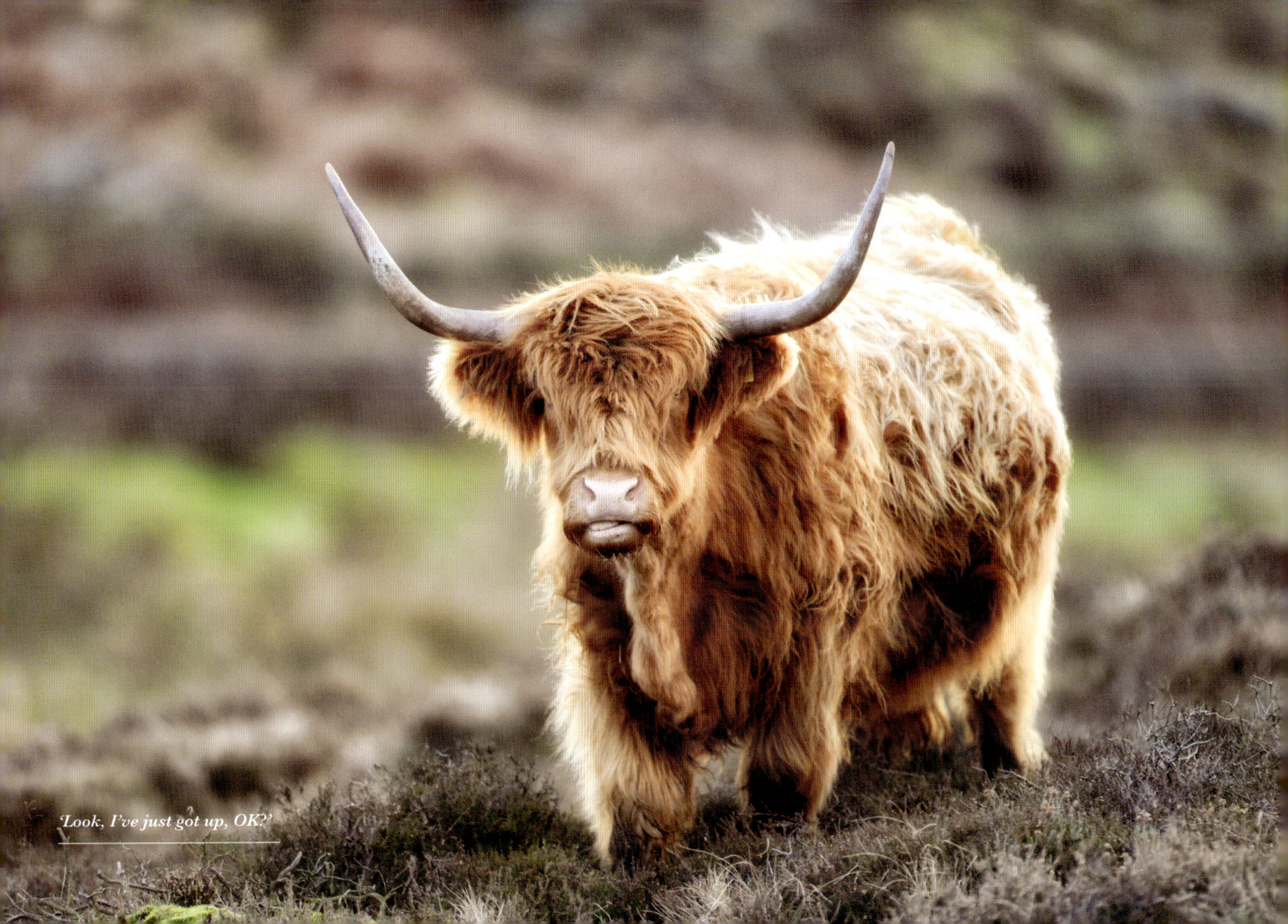

'Look, I've just got up, OK?'

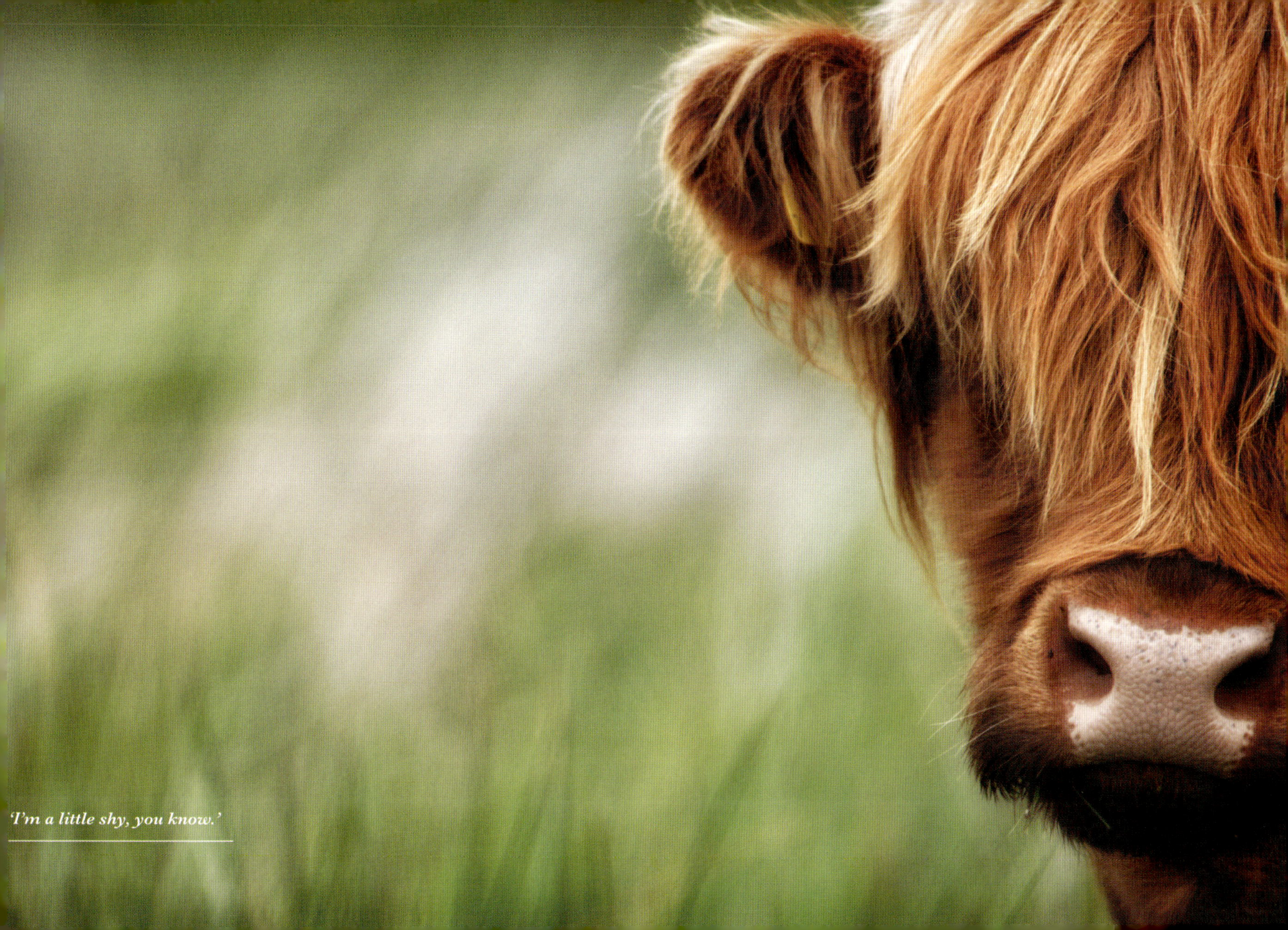
'I'm a little shy, you know.'

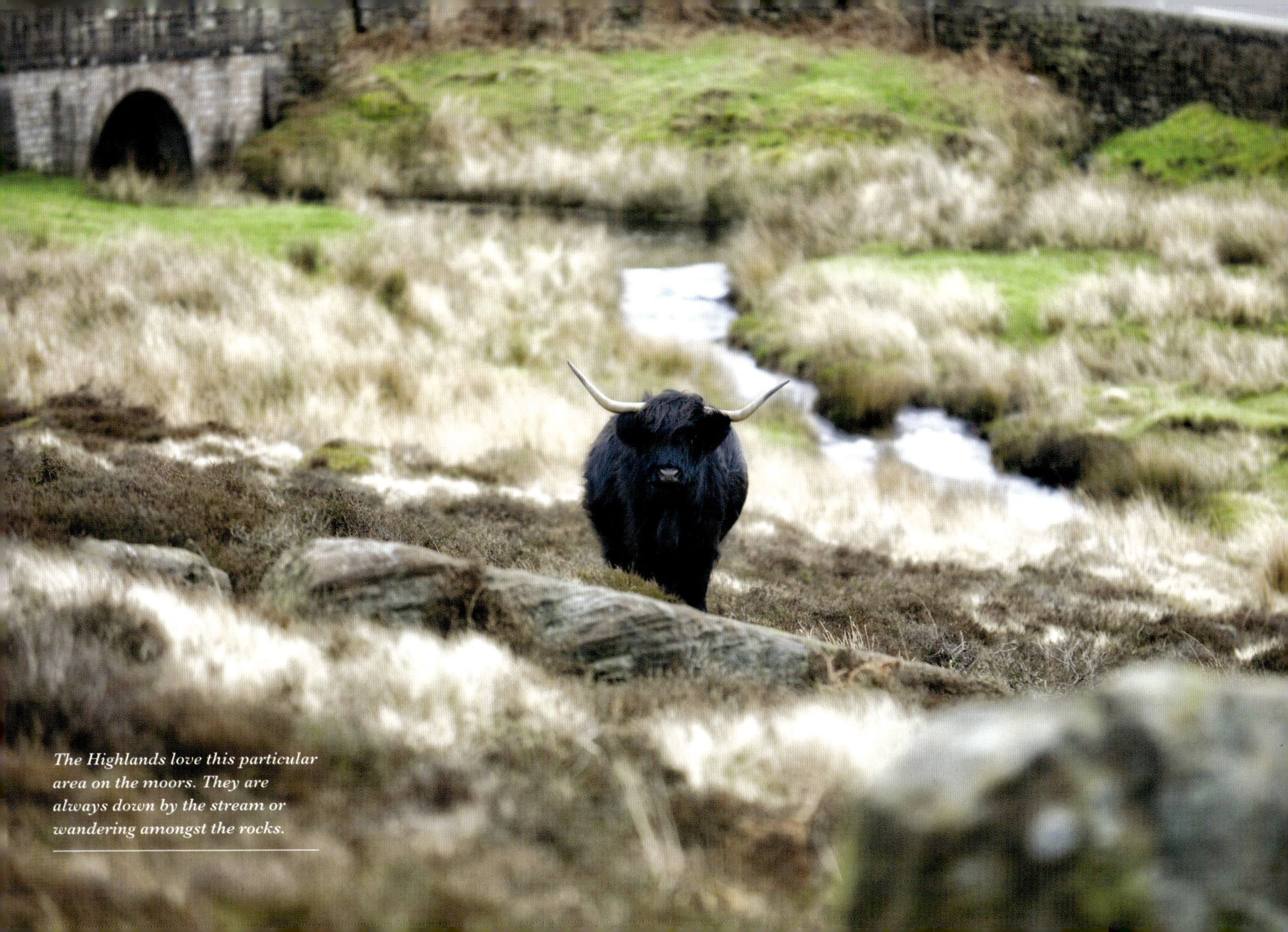

The Highlands love this particular area on the moors. They are always down by the stream or wandering amongst the rocks.

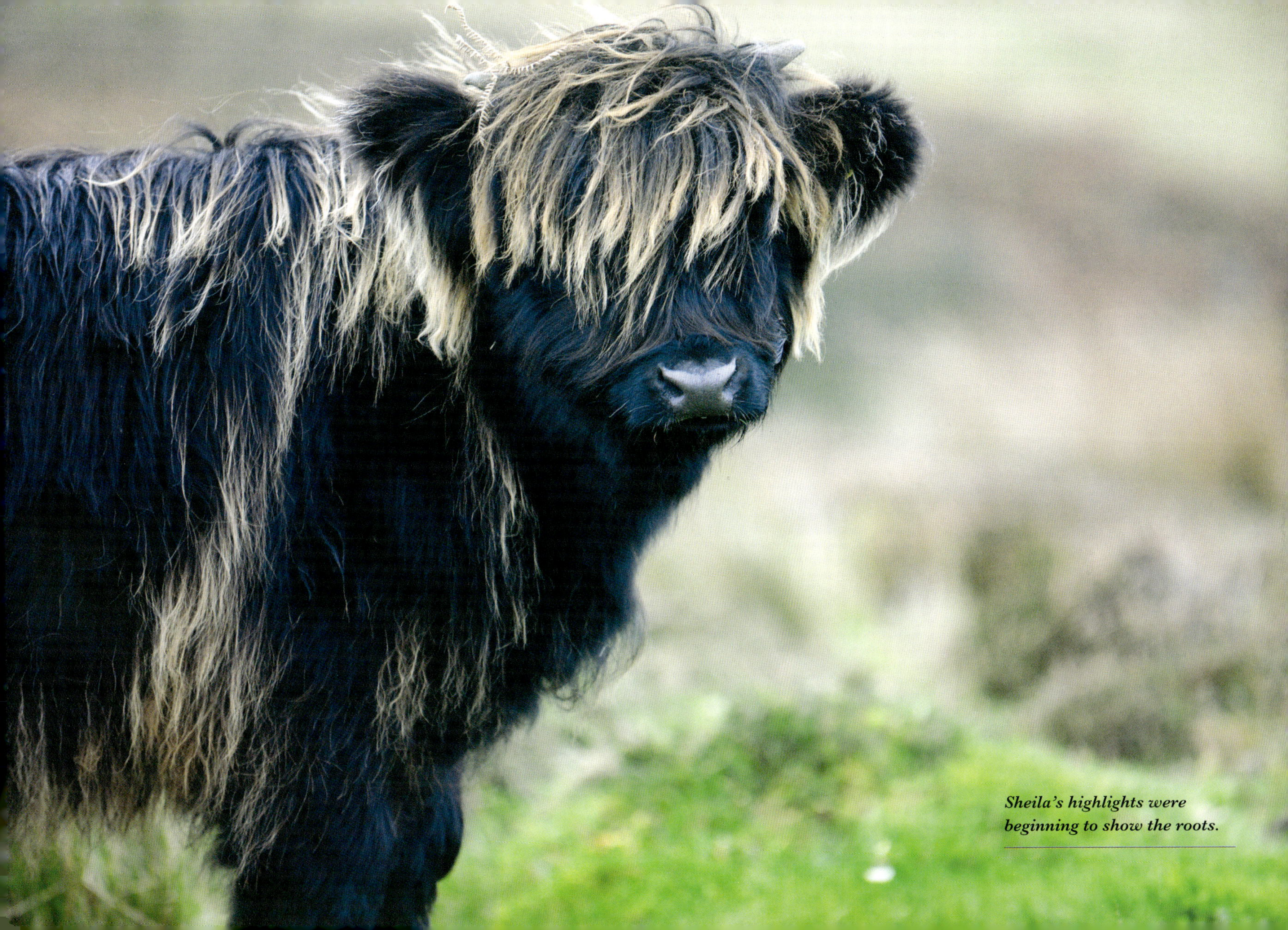

Sheila's highlights were beginning to show the roots.

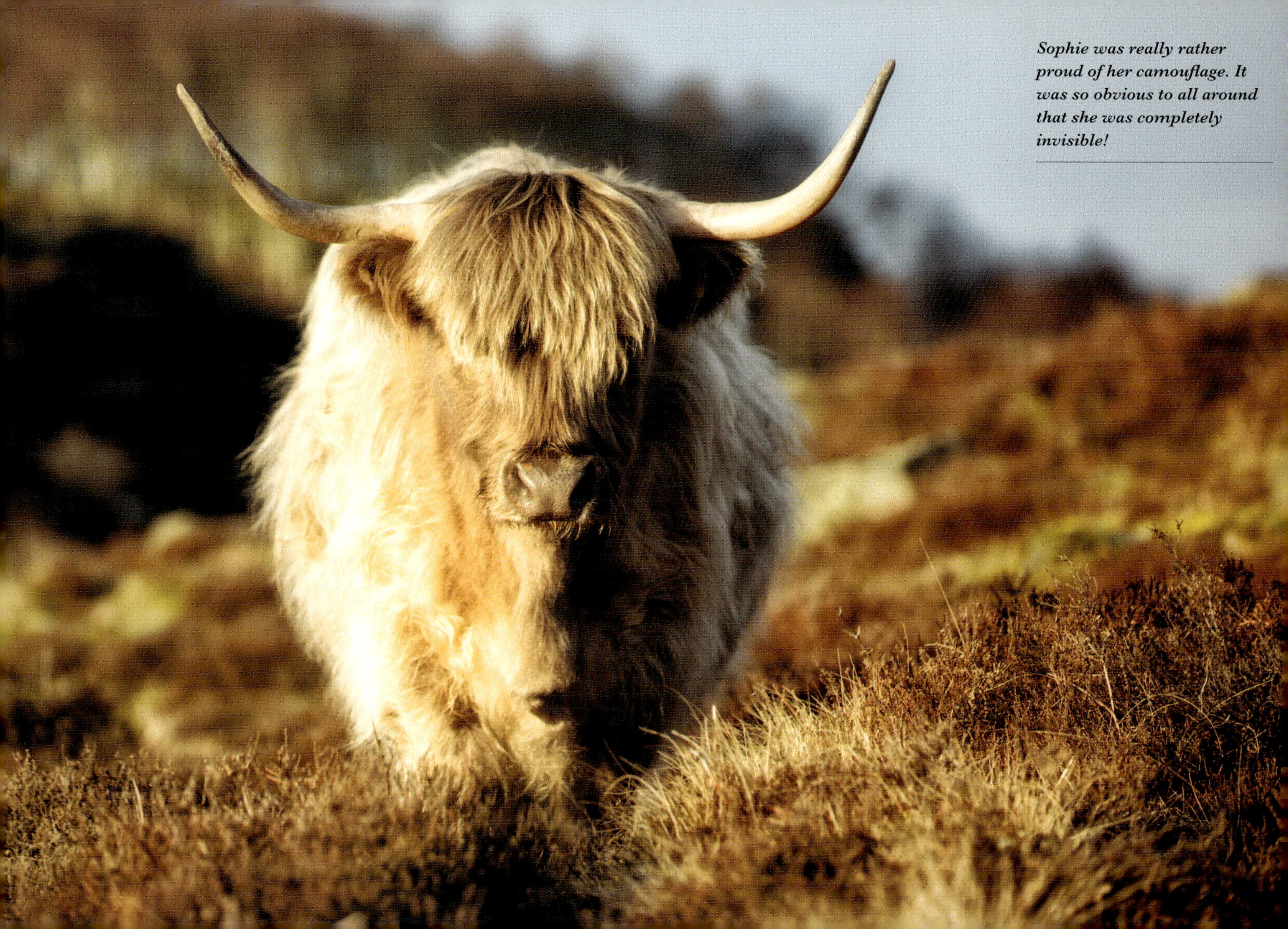
Sophie was really rather
proud of her camouflage. It
was so obvious to all around
that she was completely
invisible!

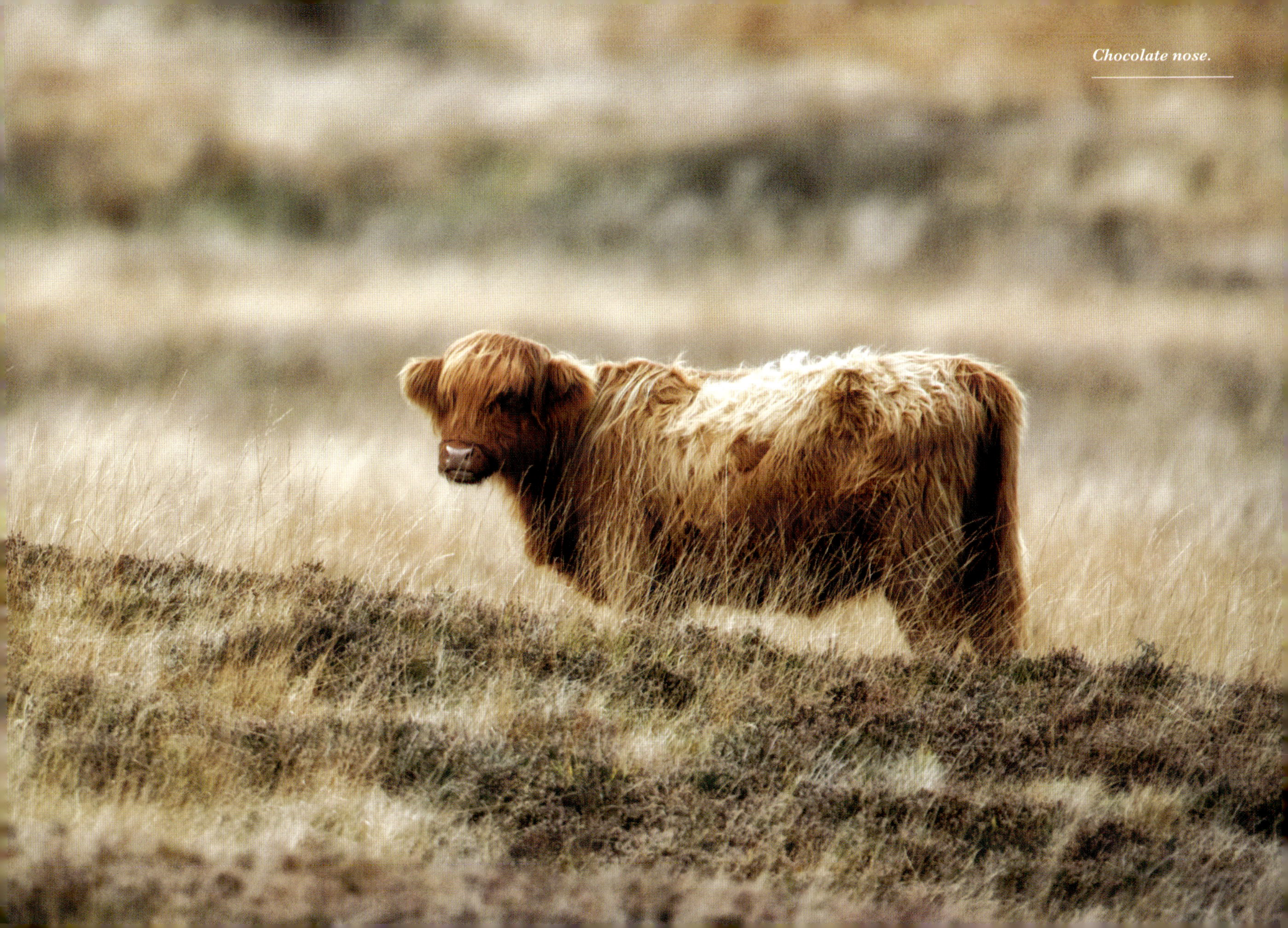
Chocolate nose.

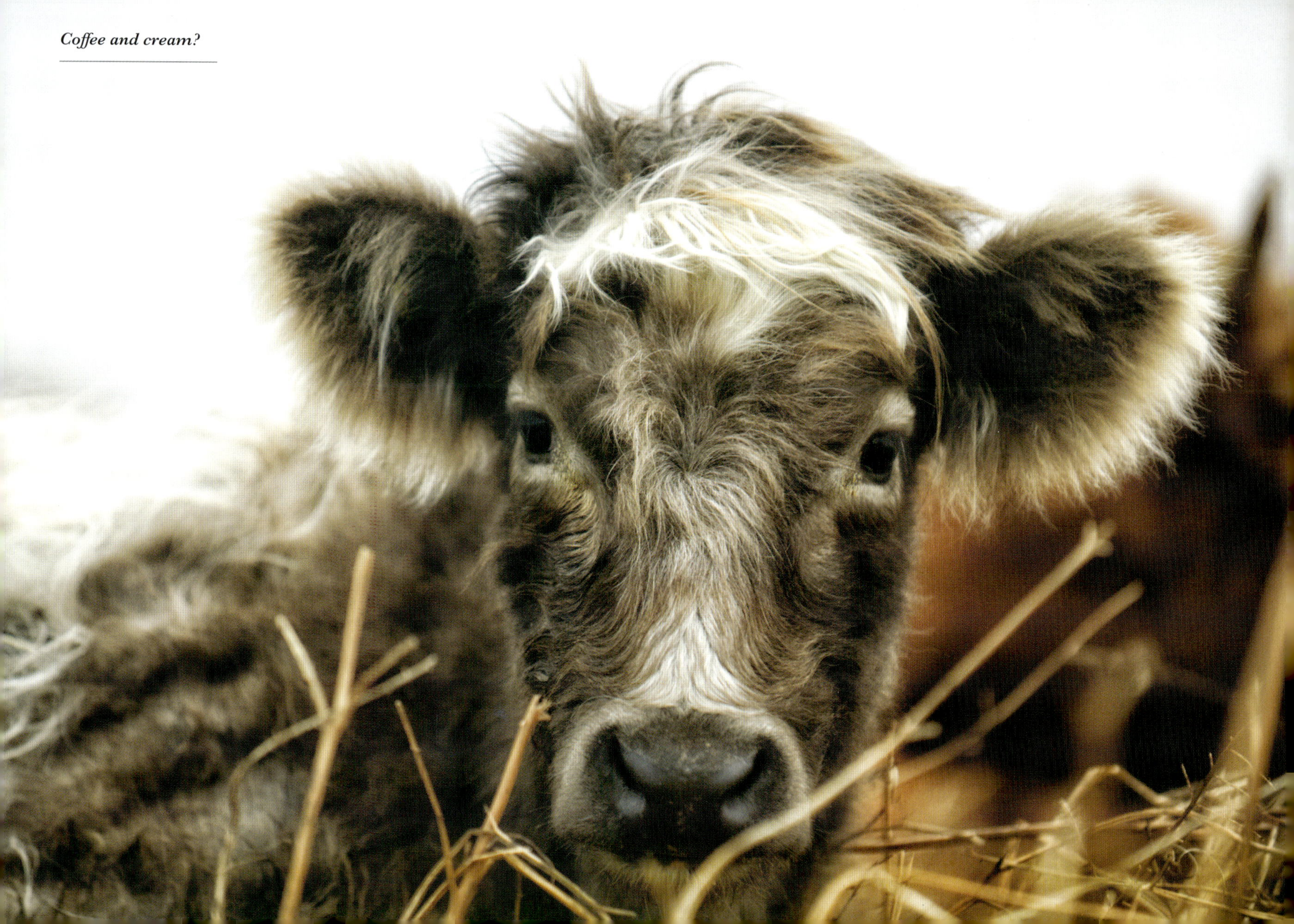

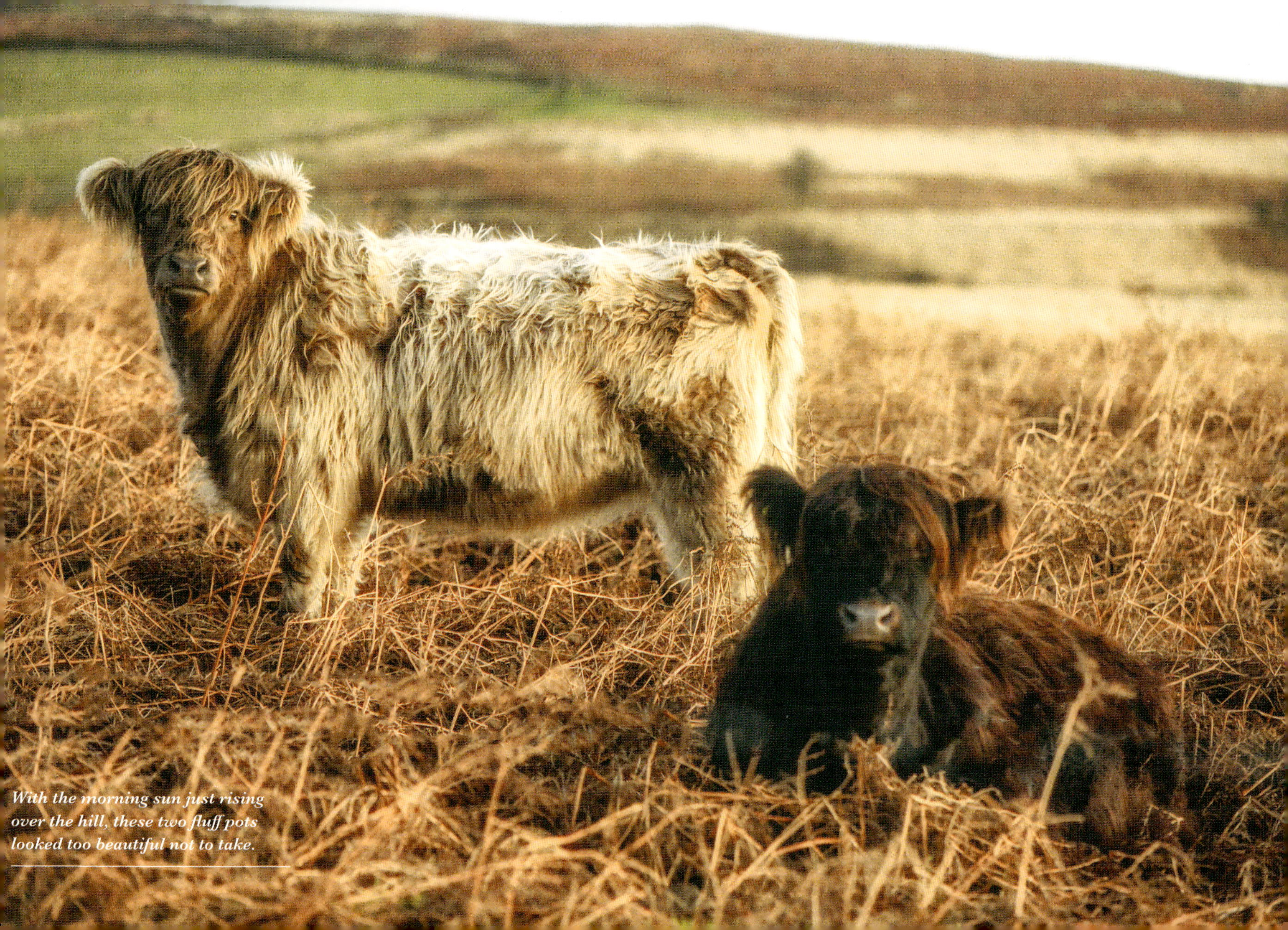

With the morning sun just rising
over the hill, these two fluff pots
looked too beautiful not to take.

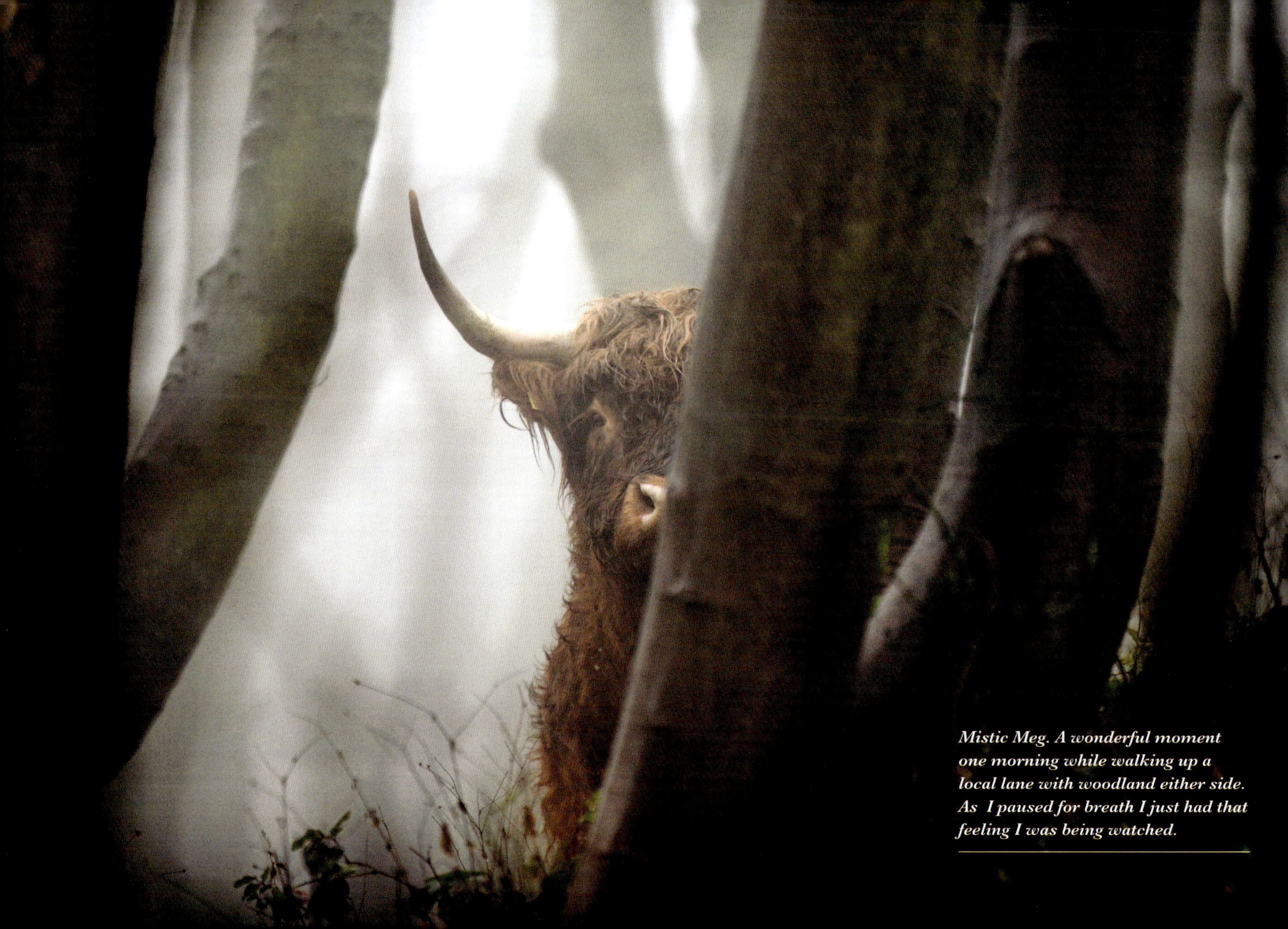

Mistic Meg. A wonderful moment one morning while walking up a local lane with woodland either side. As I paused for breath I just had that feeling I was being watched.

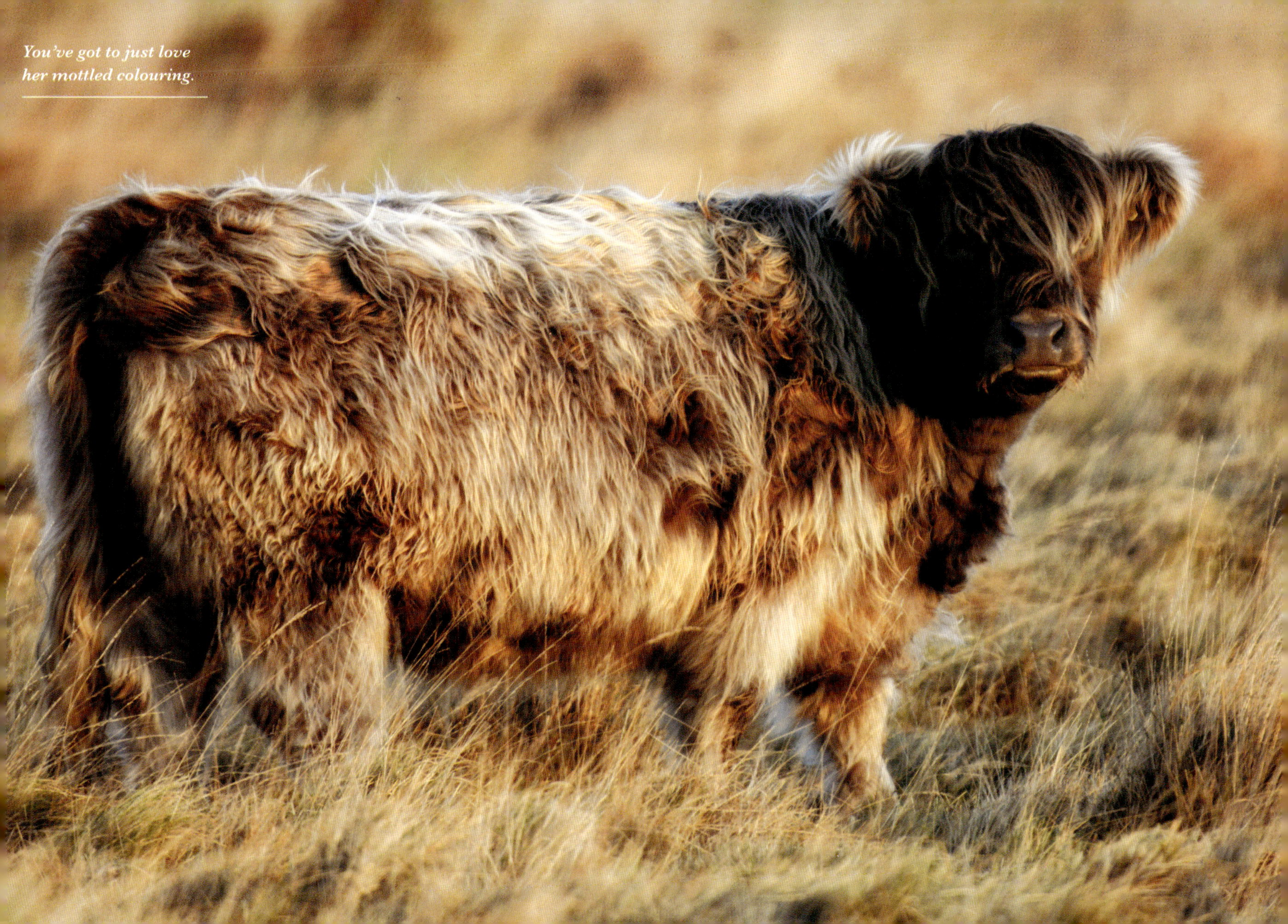
You've got to just love
her mottled colouring.

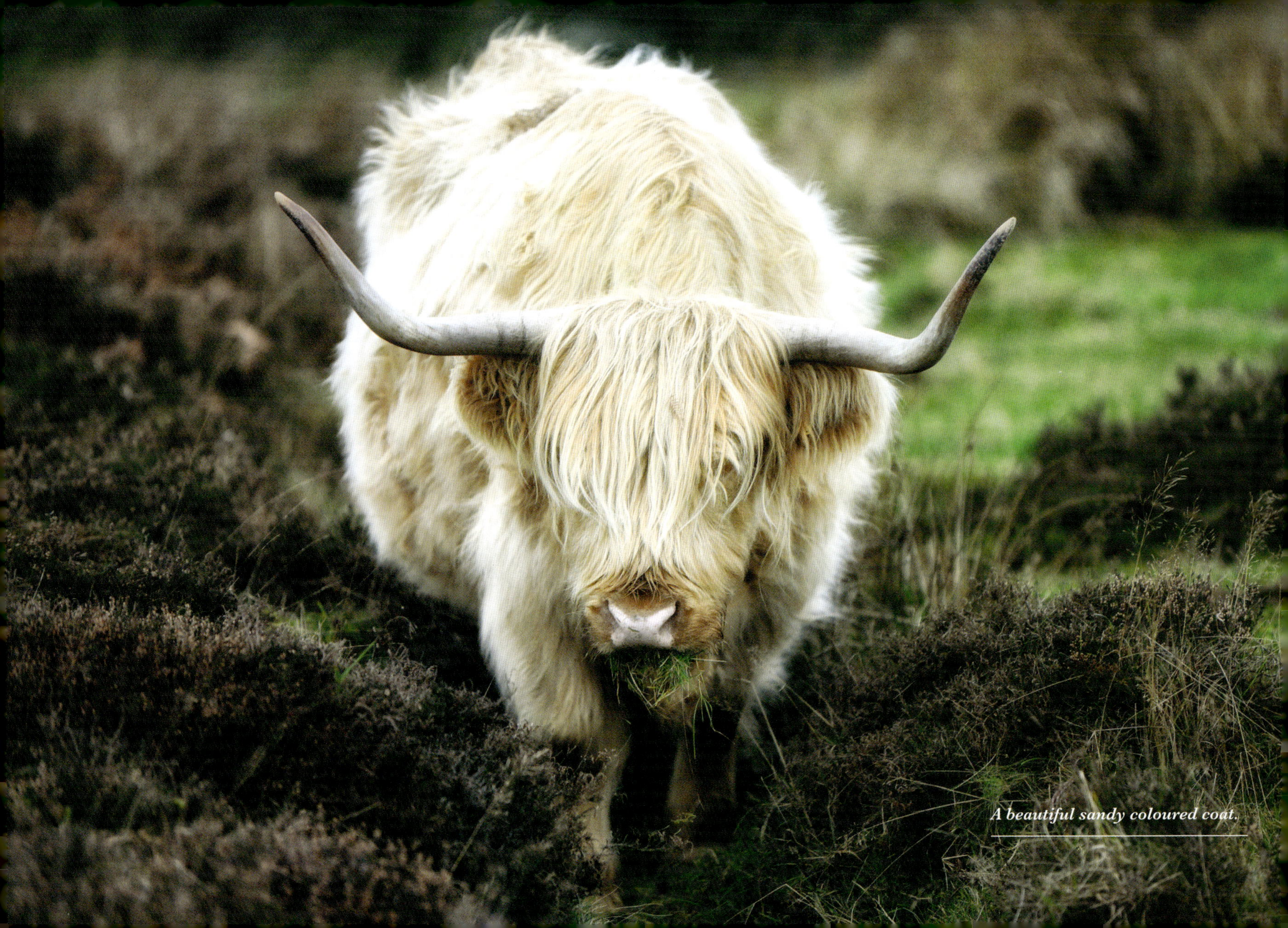

A beautiful sandy coloured coat.

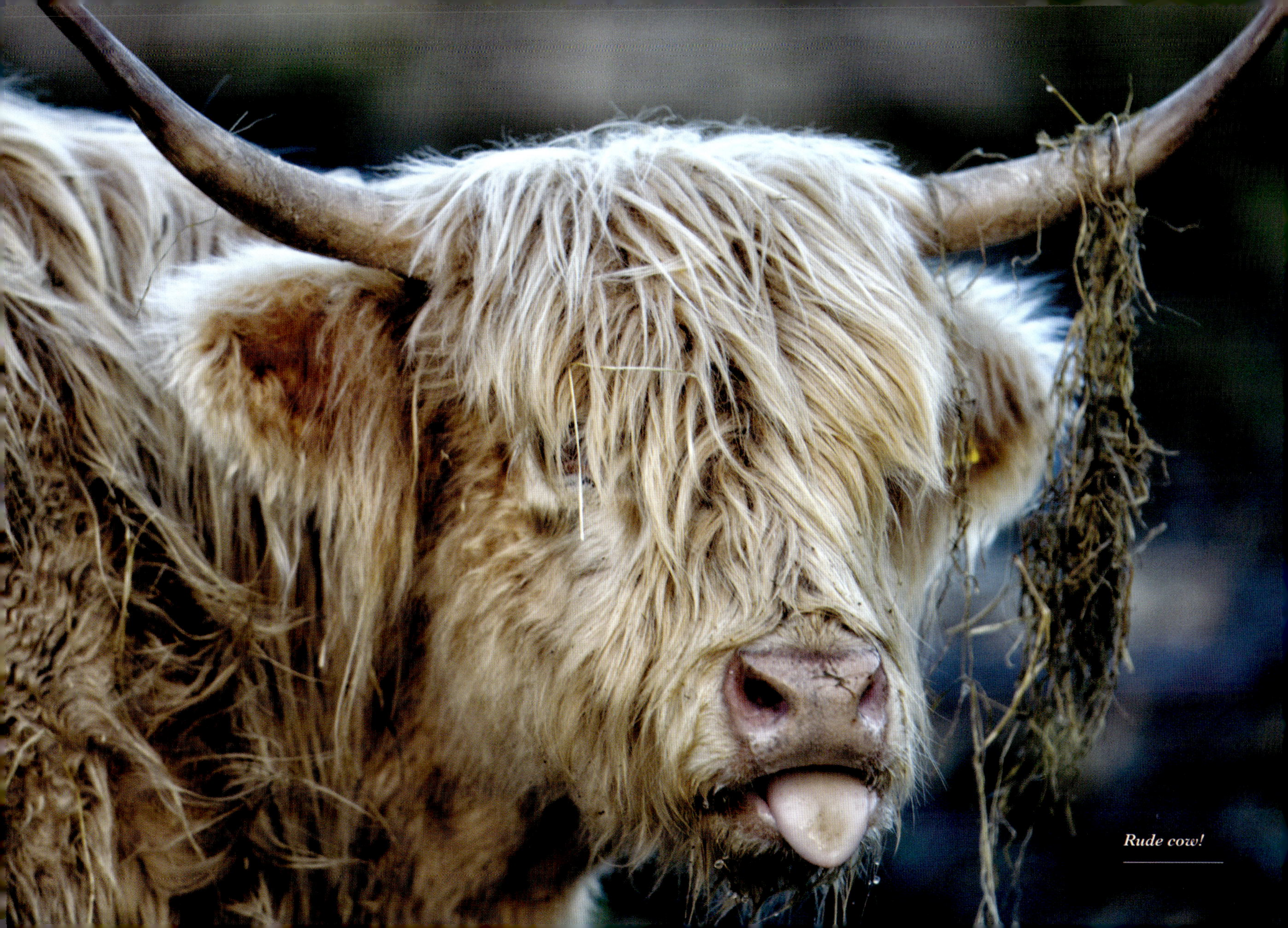

Rude cow!

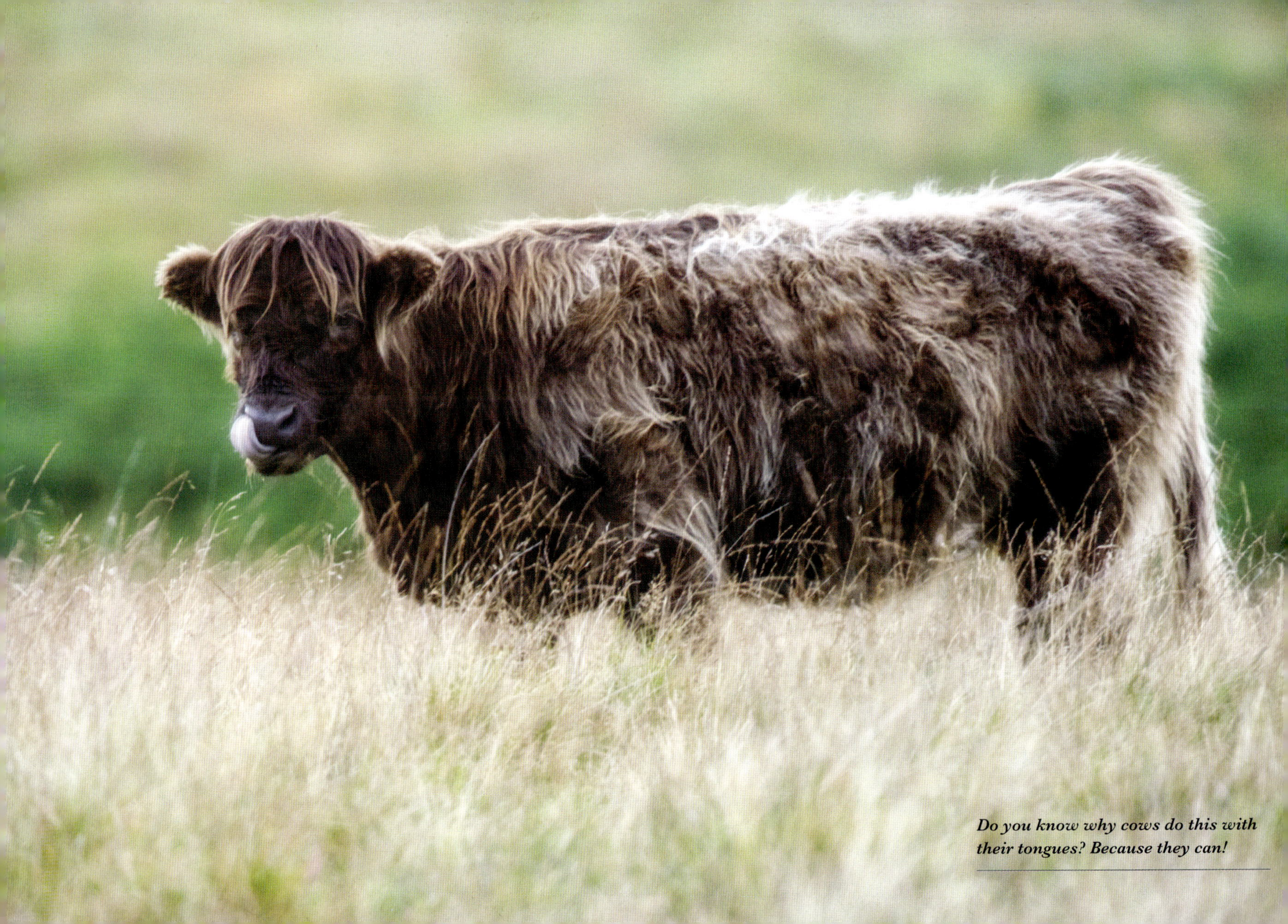

Do you know why cows do this with their tongues? Because they can!

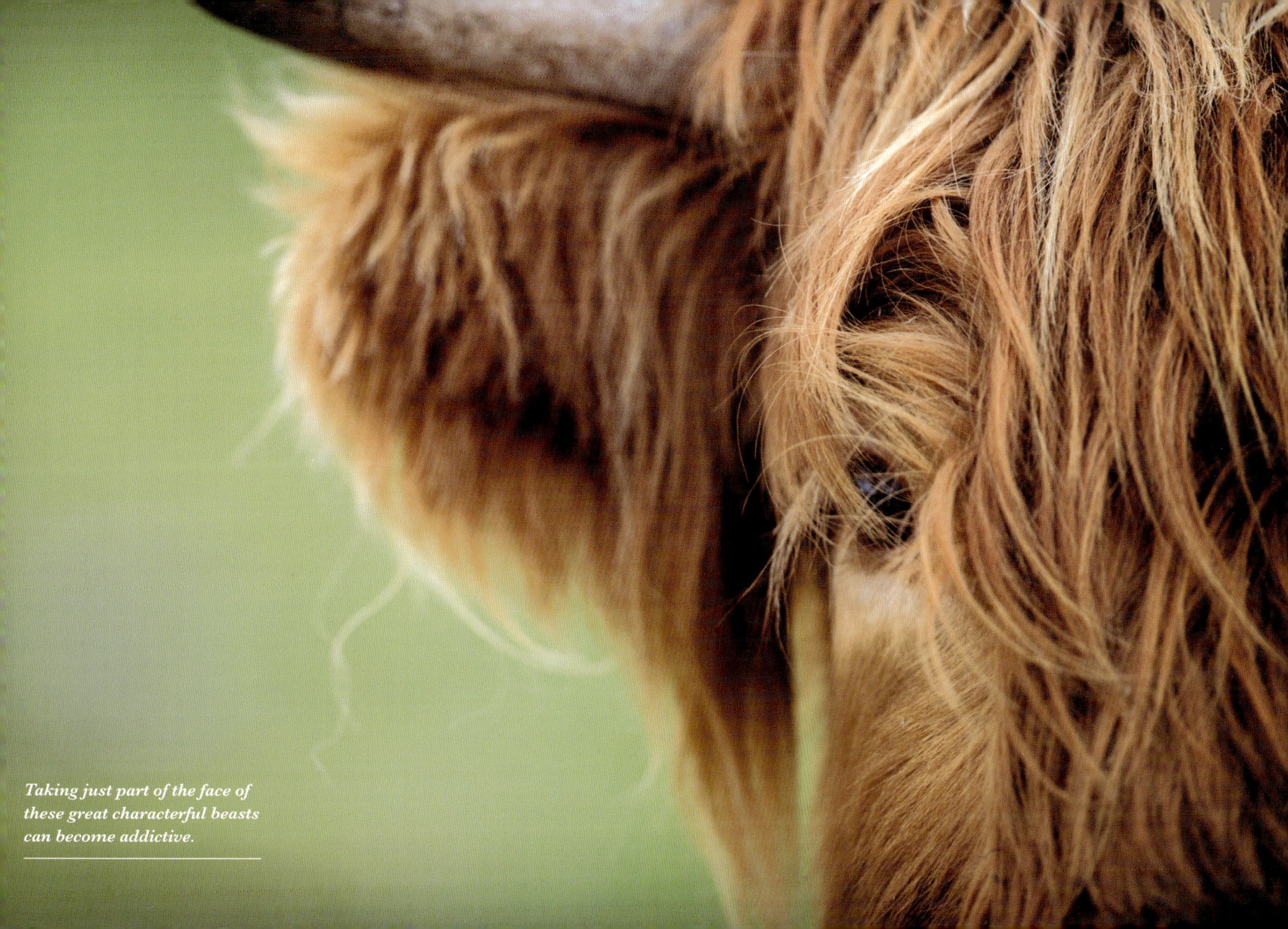
Taking just part of the face of
these great characterful beasts
can become addictive.

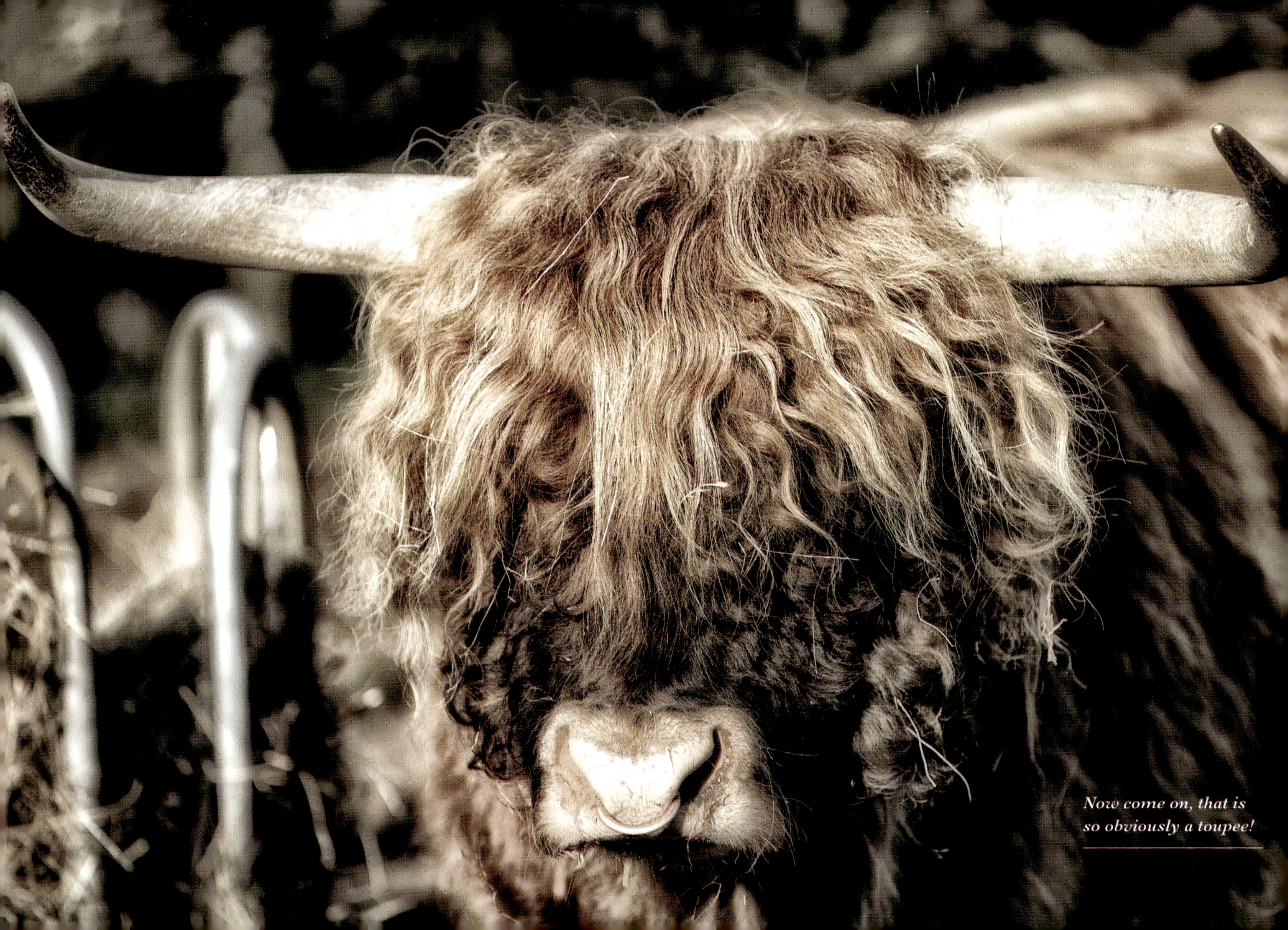
Now come on, that is
so obviously a toupee!

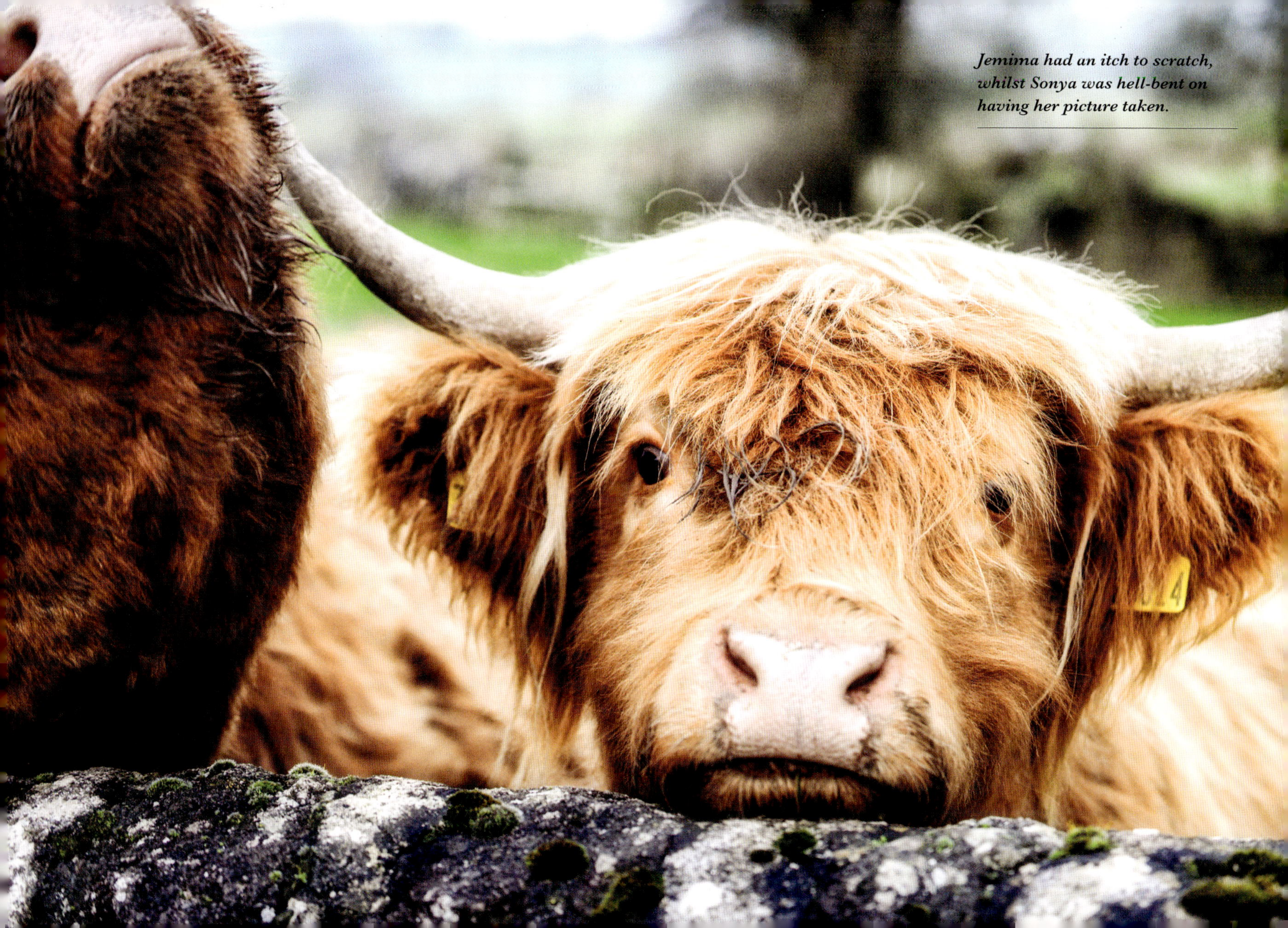

Jemima had an itch to scratch, whilst Sonya was hell-bent on having her picture taken.

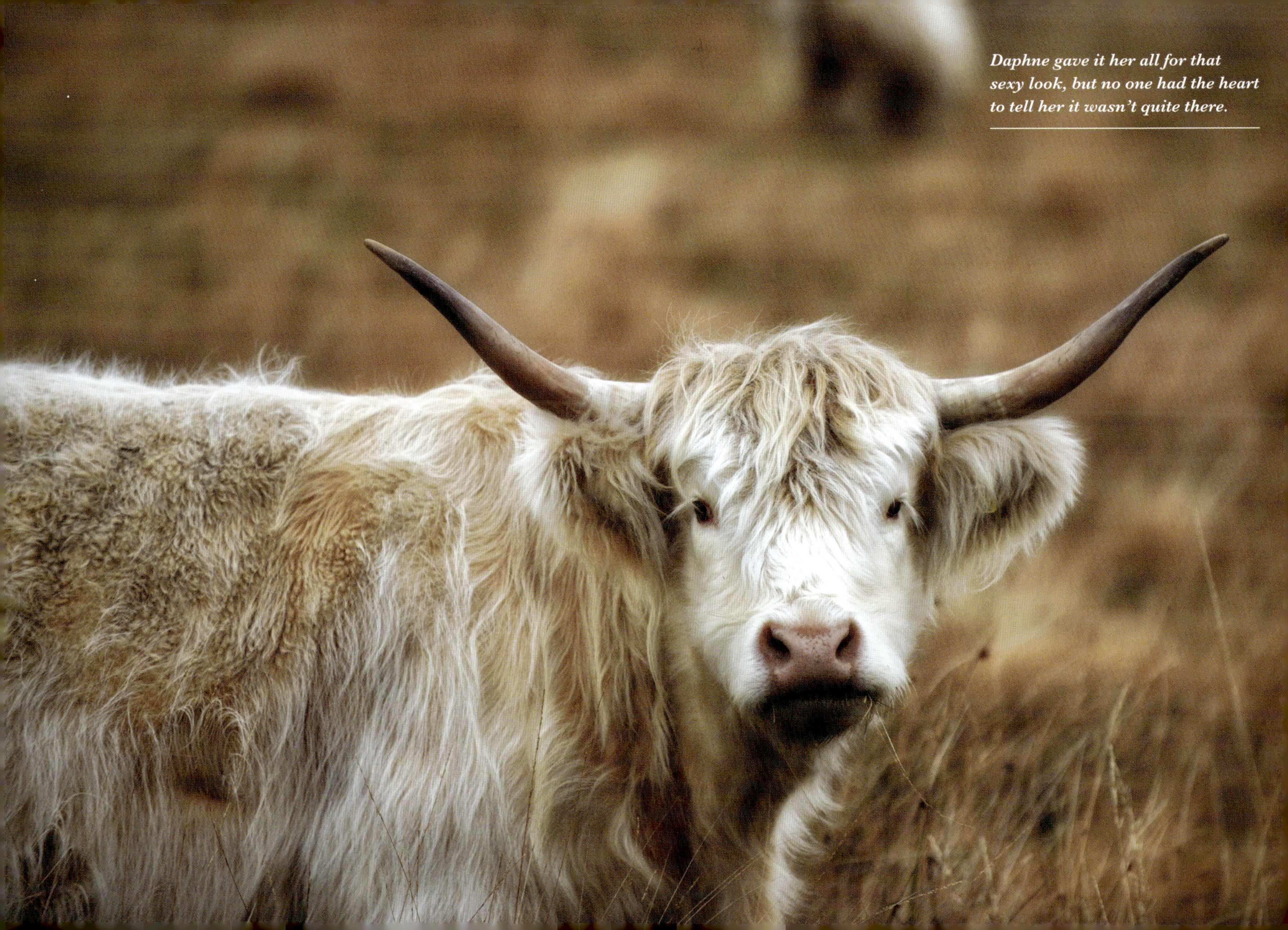
Daphne gave it her all for that sexy look, but no one had the heart to tell her it wasn't quite there.

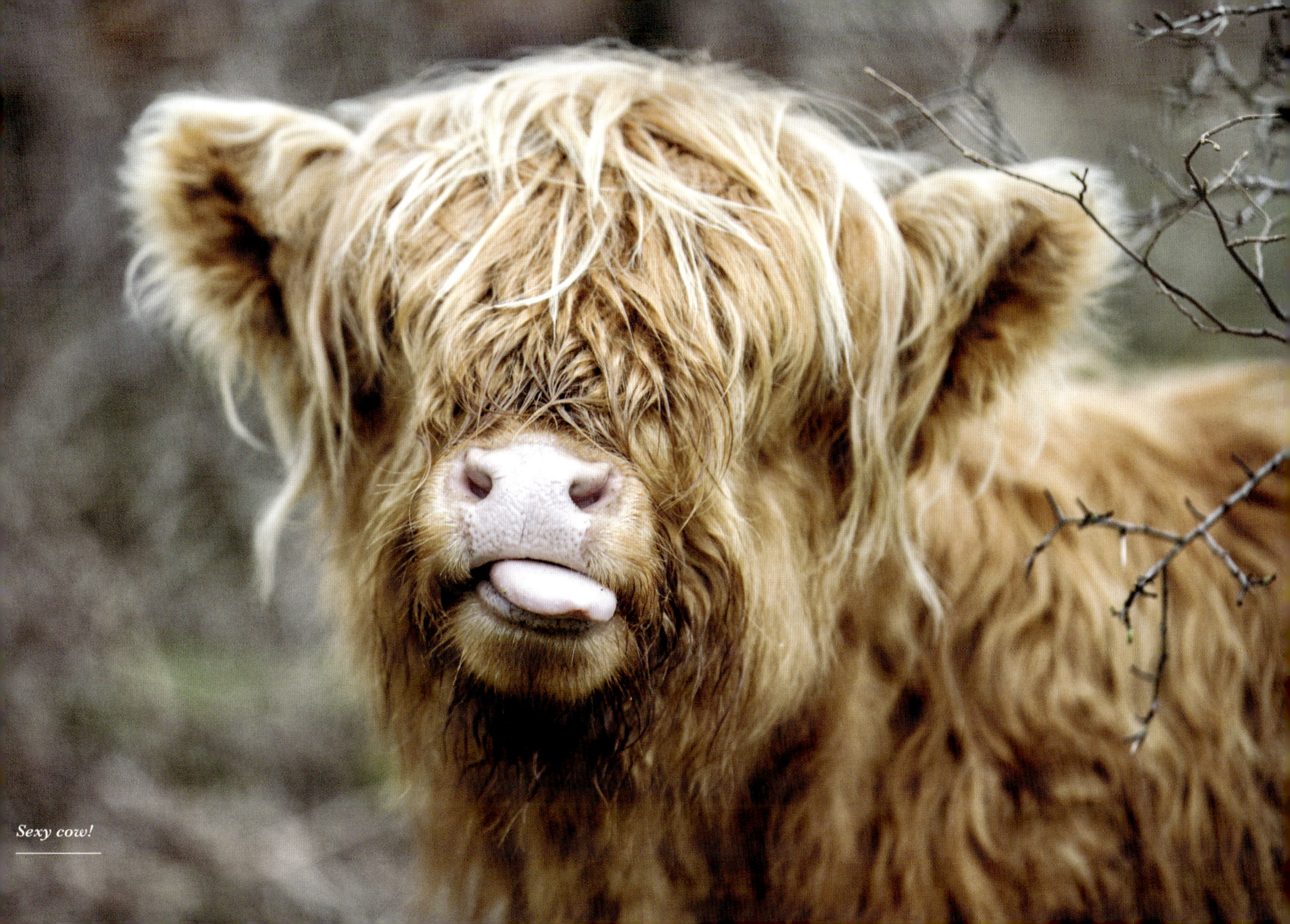

Sexy cow!

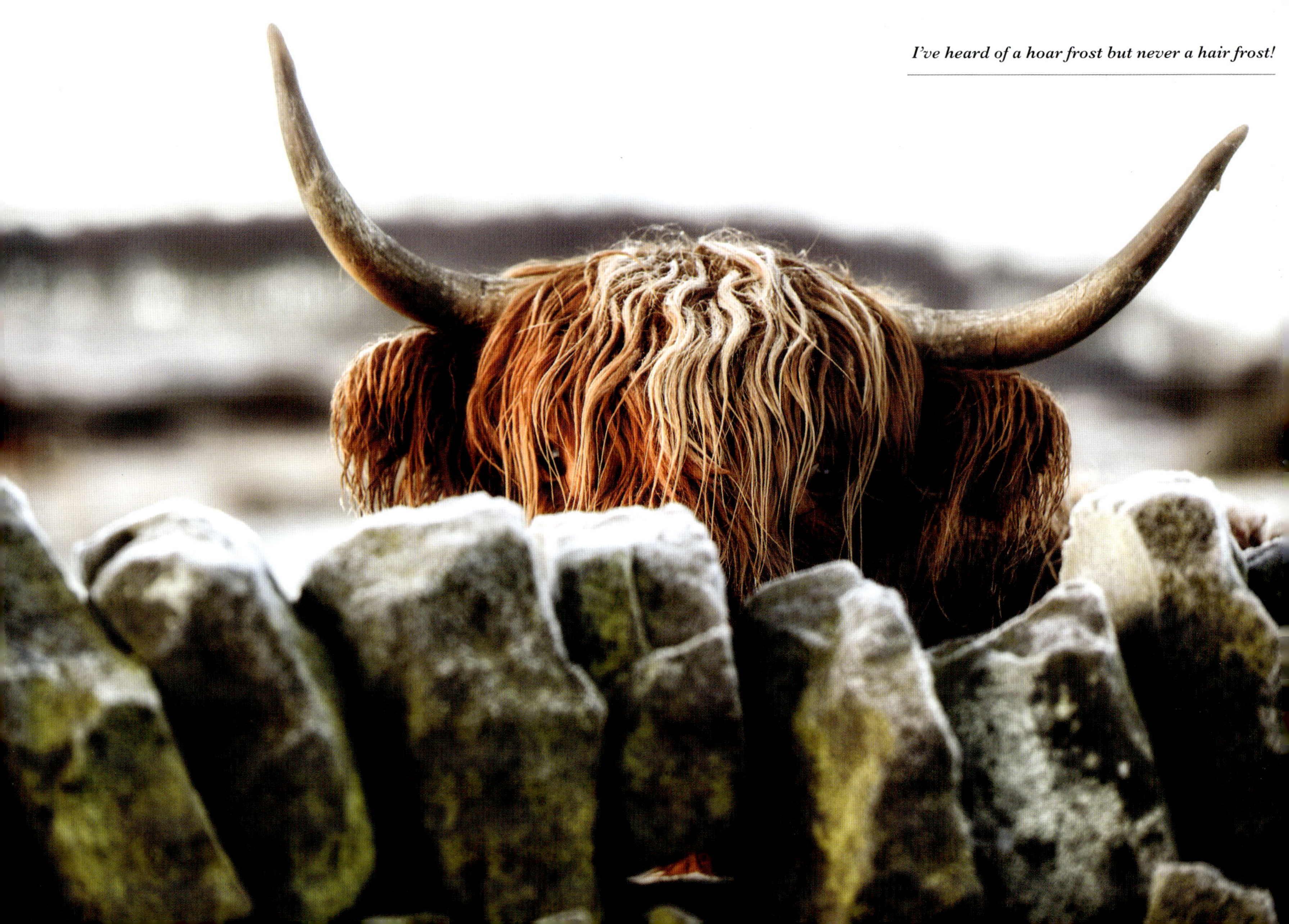

I've heard of a hoar frost but never a hair frost!

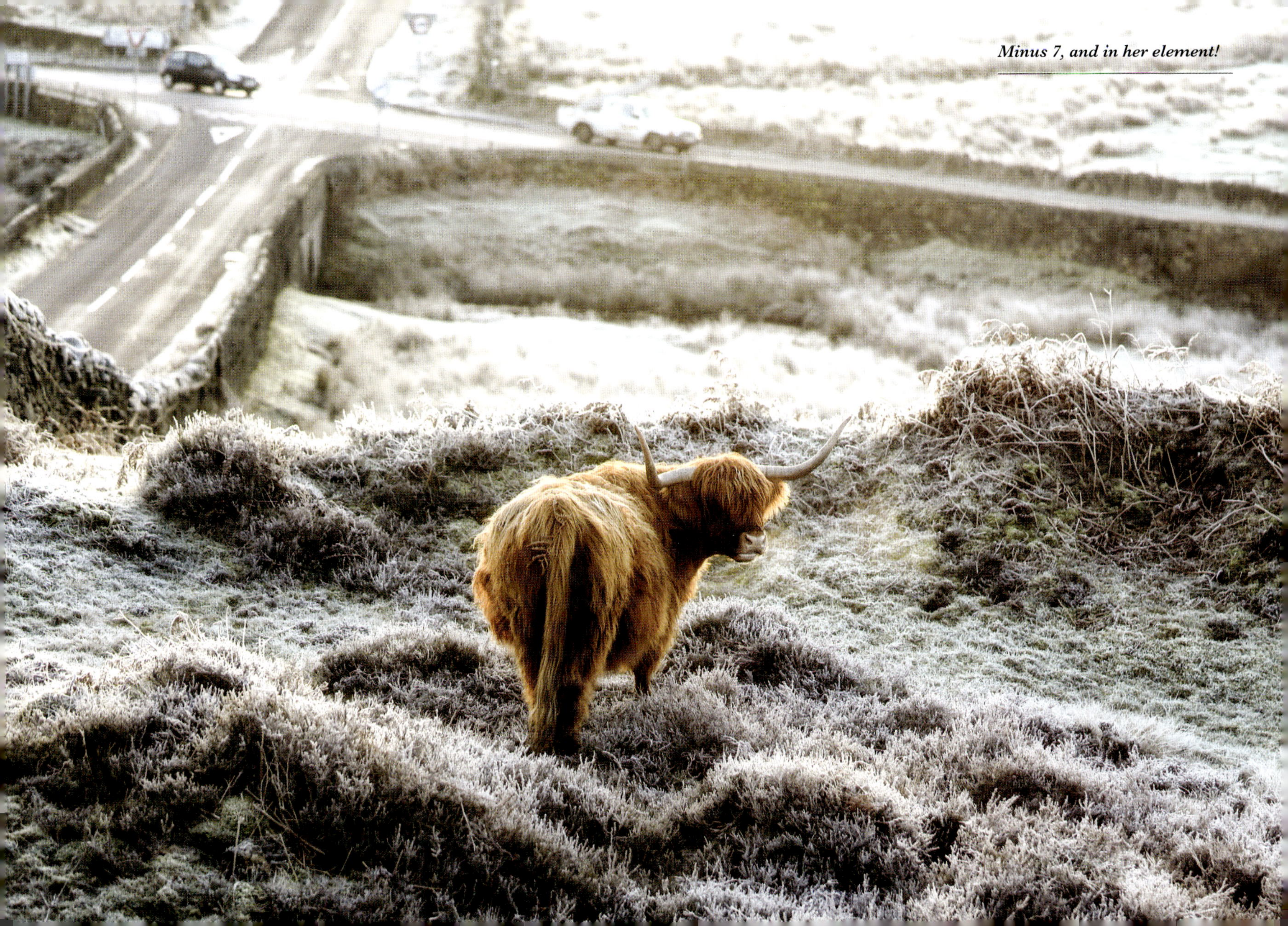

Minus 7, and in her element!

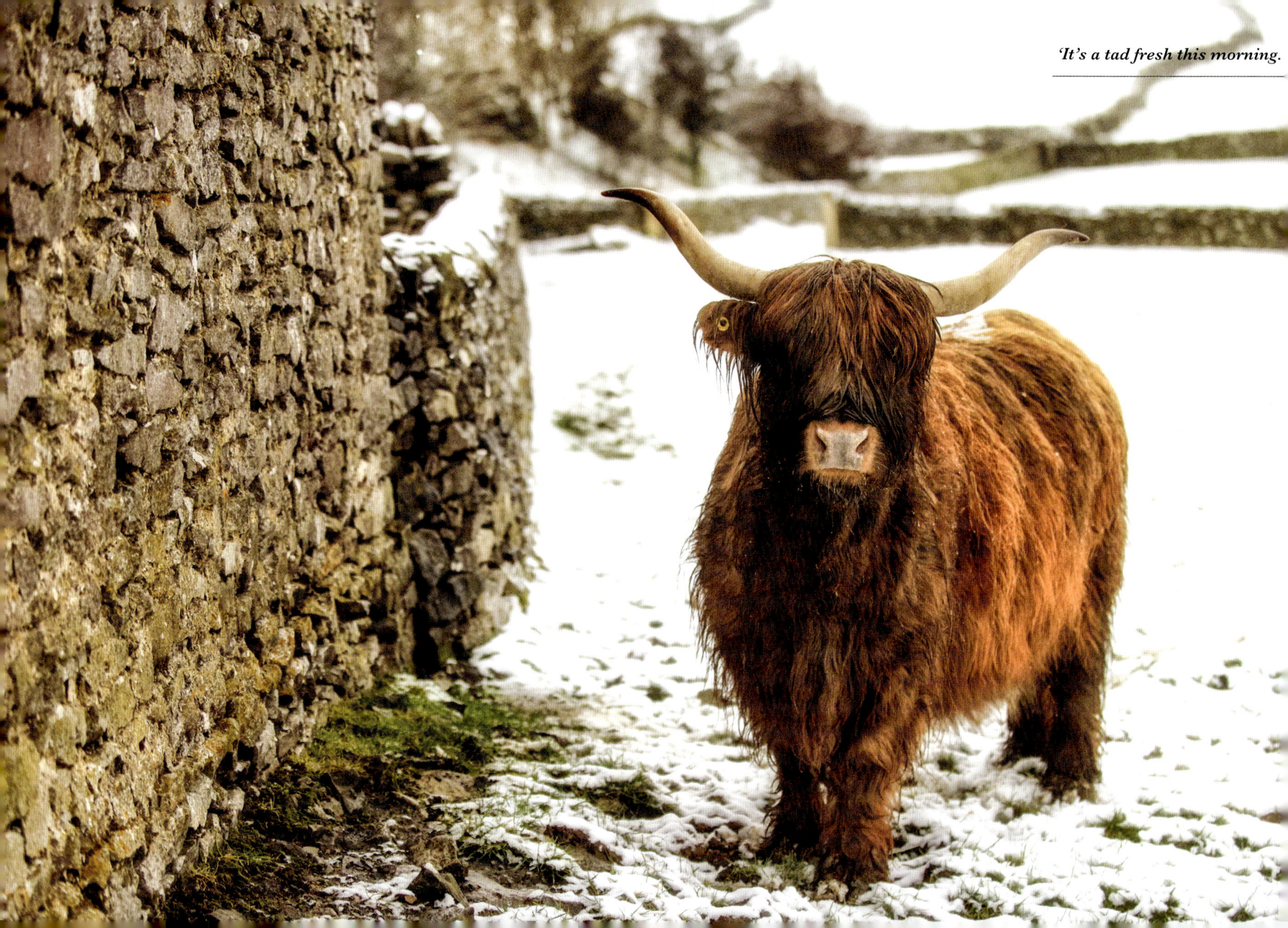

'It's a tad fresh this morning.

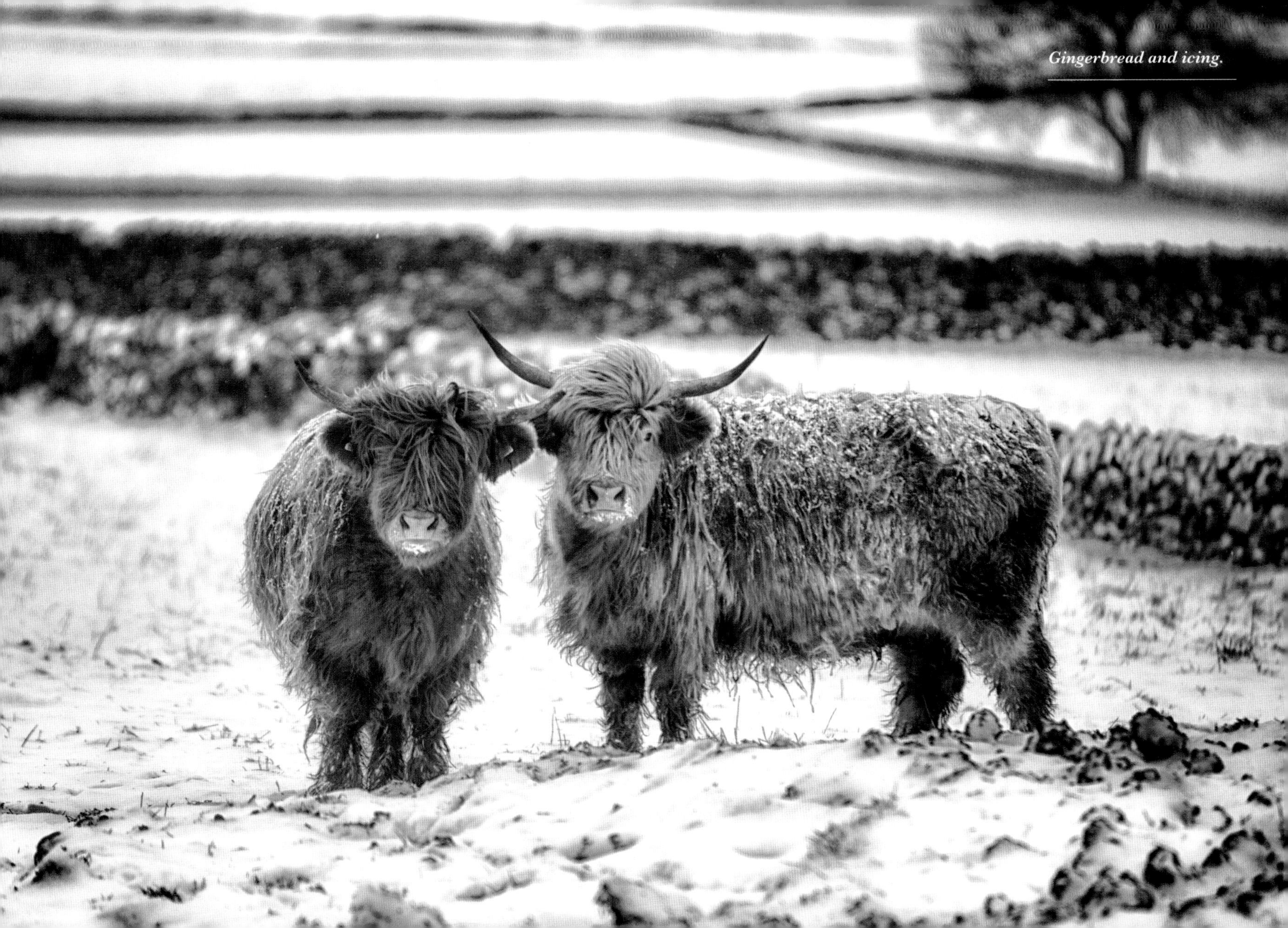
Gingerbread and icing.

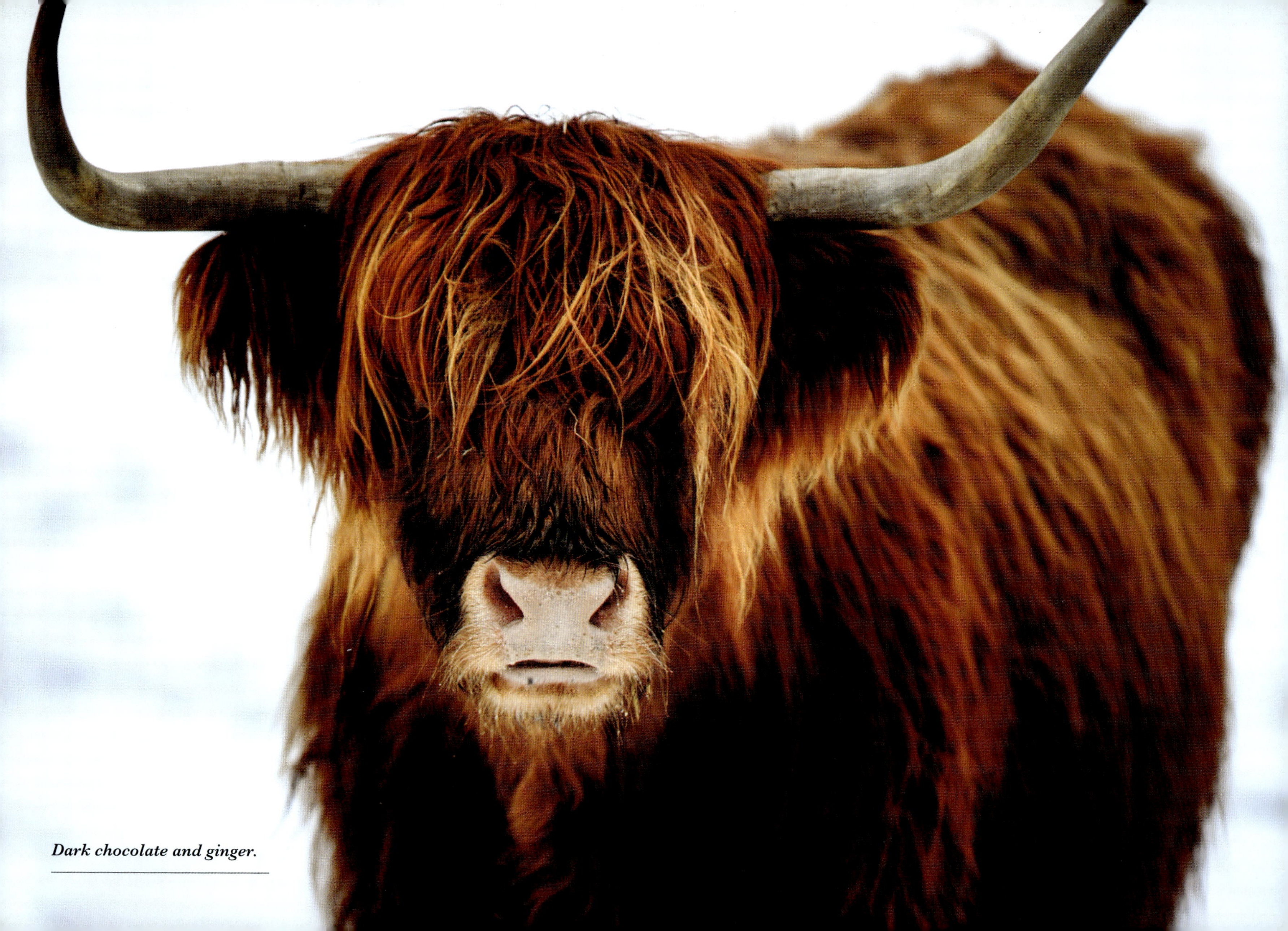

Dark chocolate and ginger.

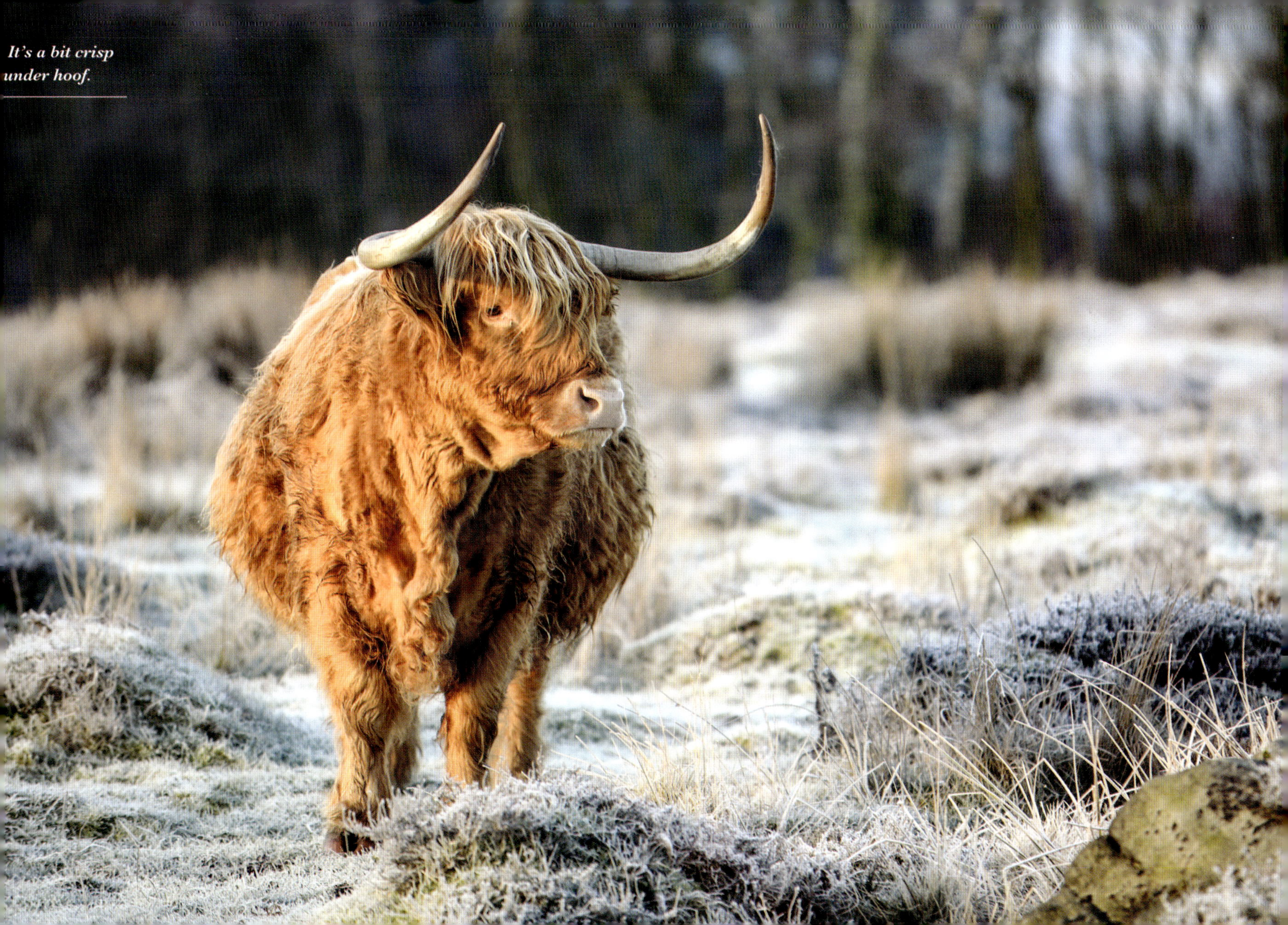
It's a bit crisp
under hoof.

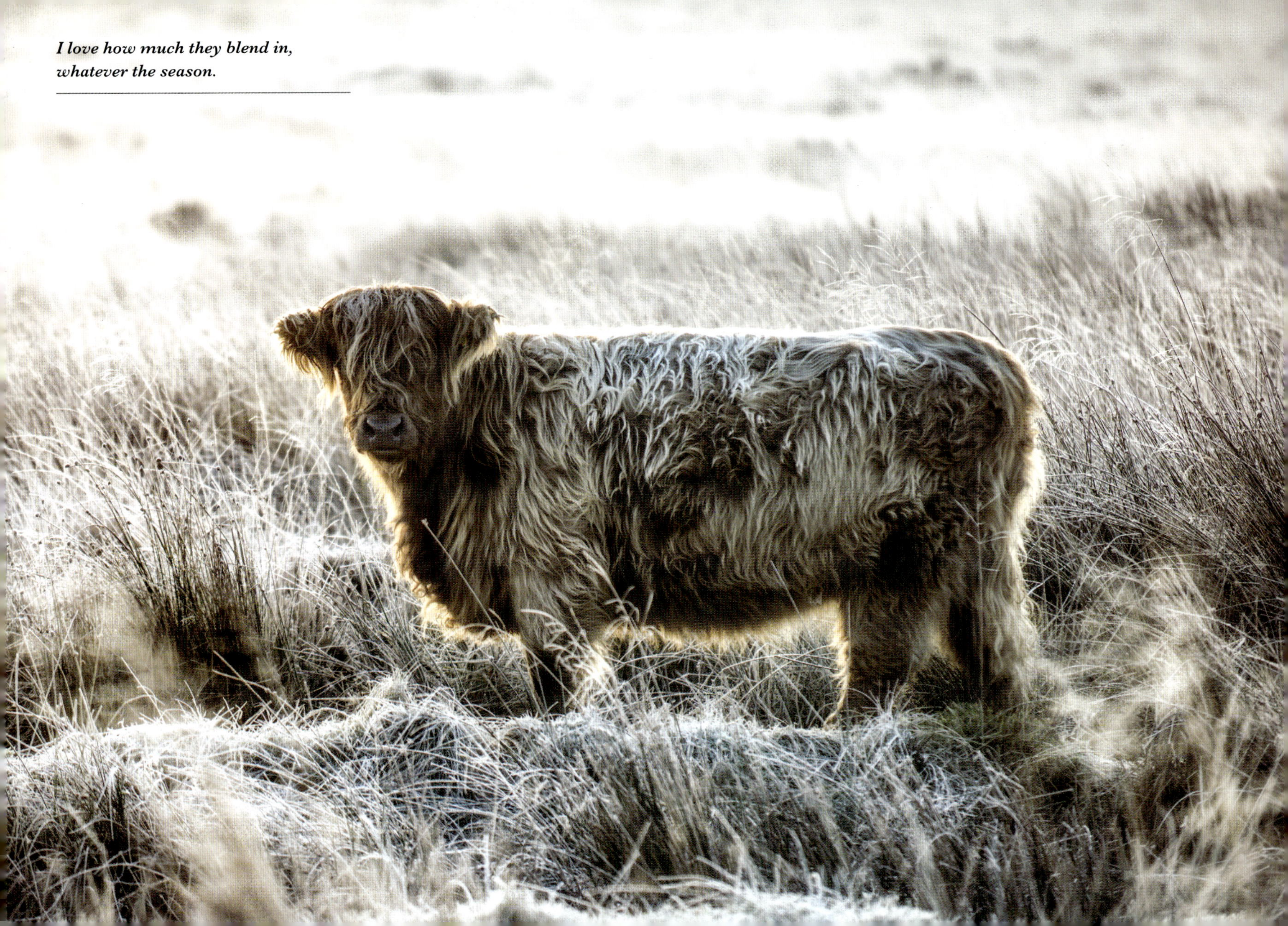
I love how much they blend in,
whatever the season.

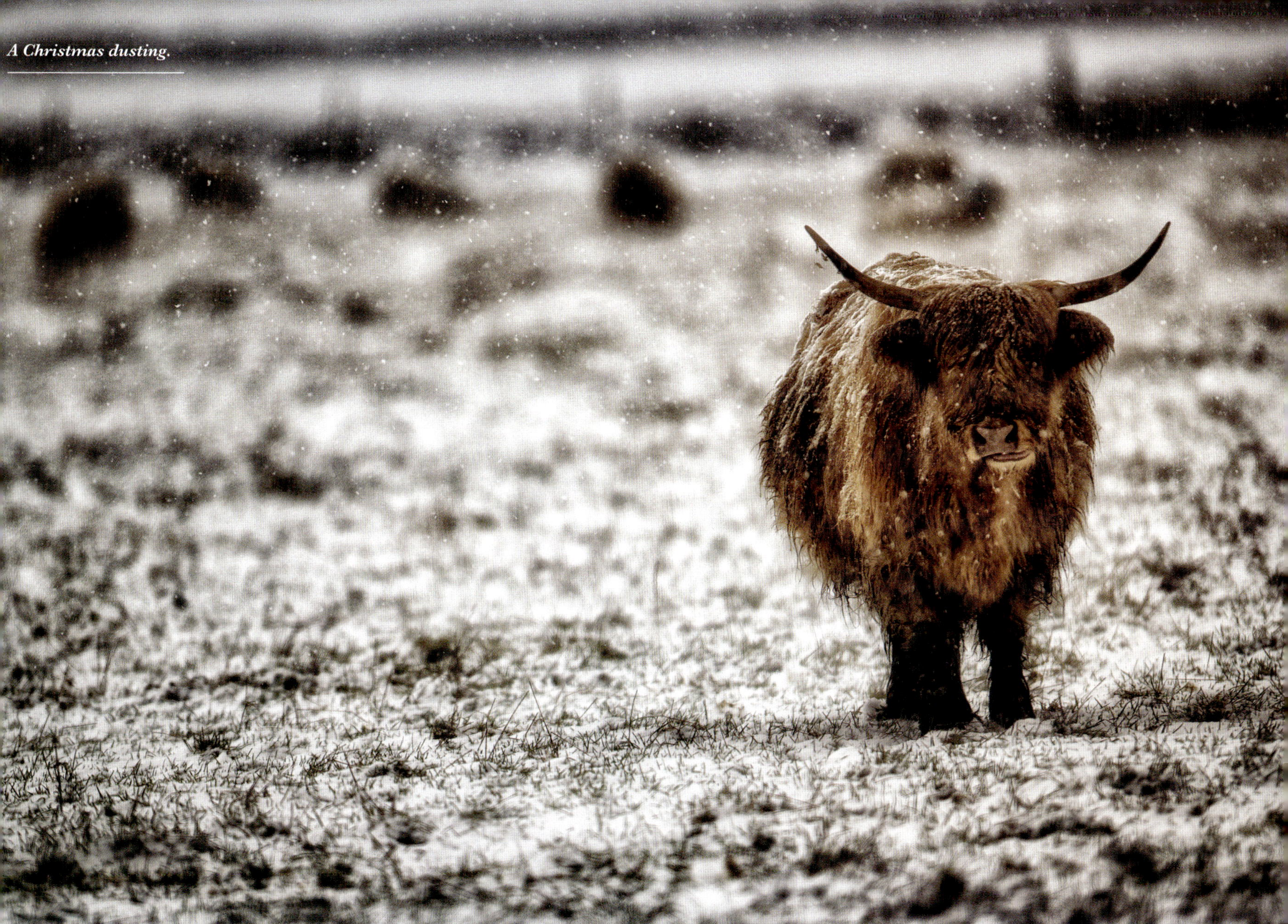

A Christmas dusting.

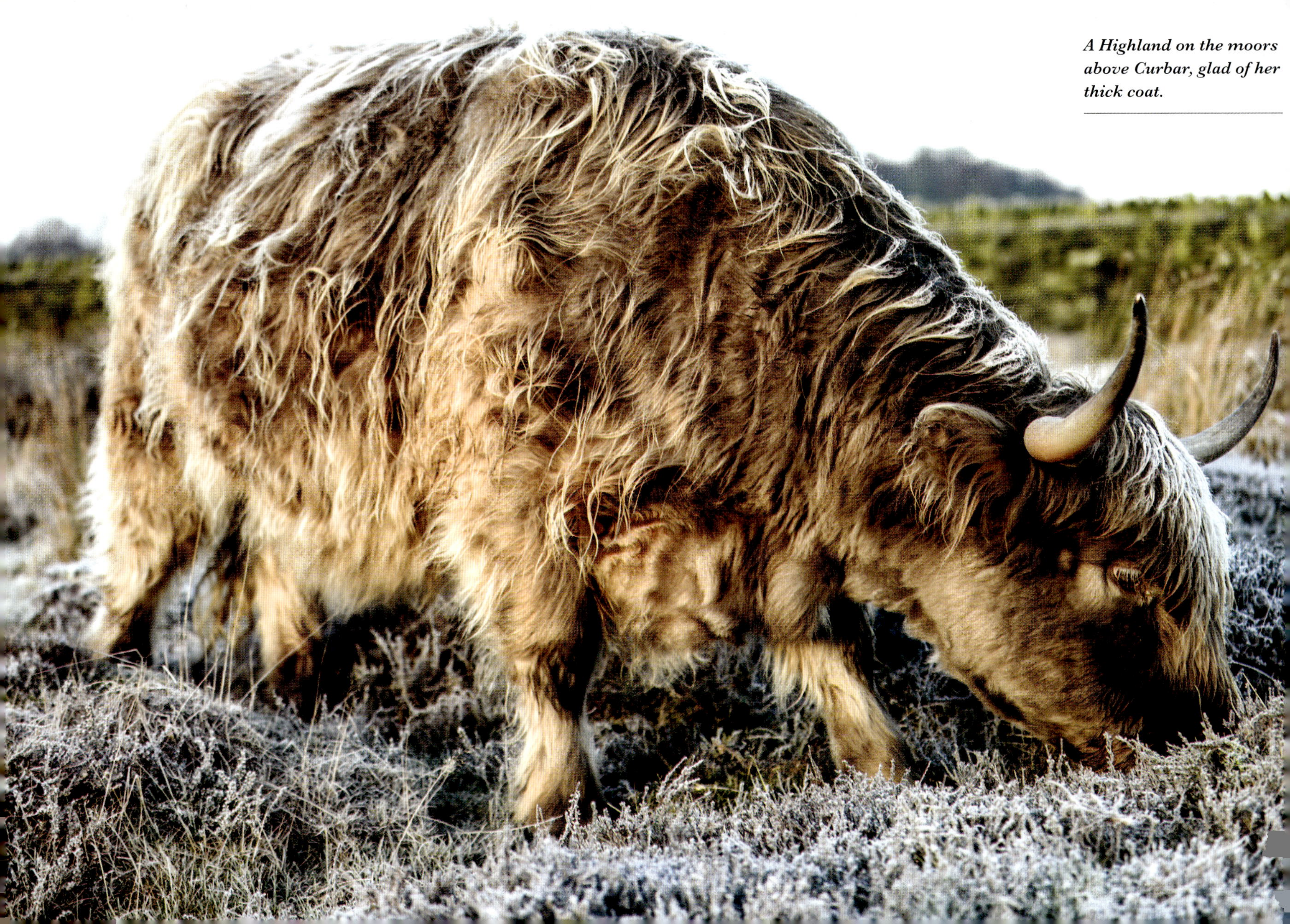

A Highland on the moors
above Curbar, glad of her
thick coat.

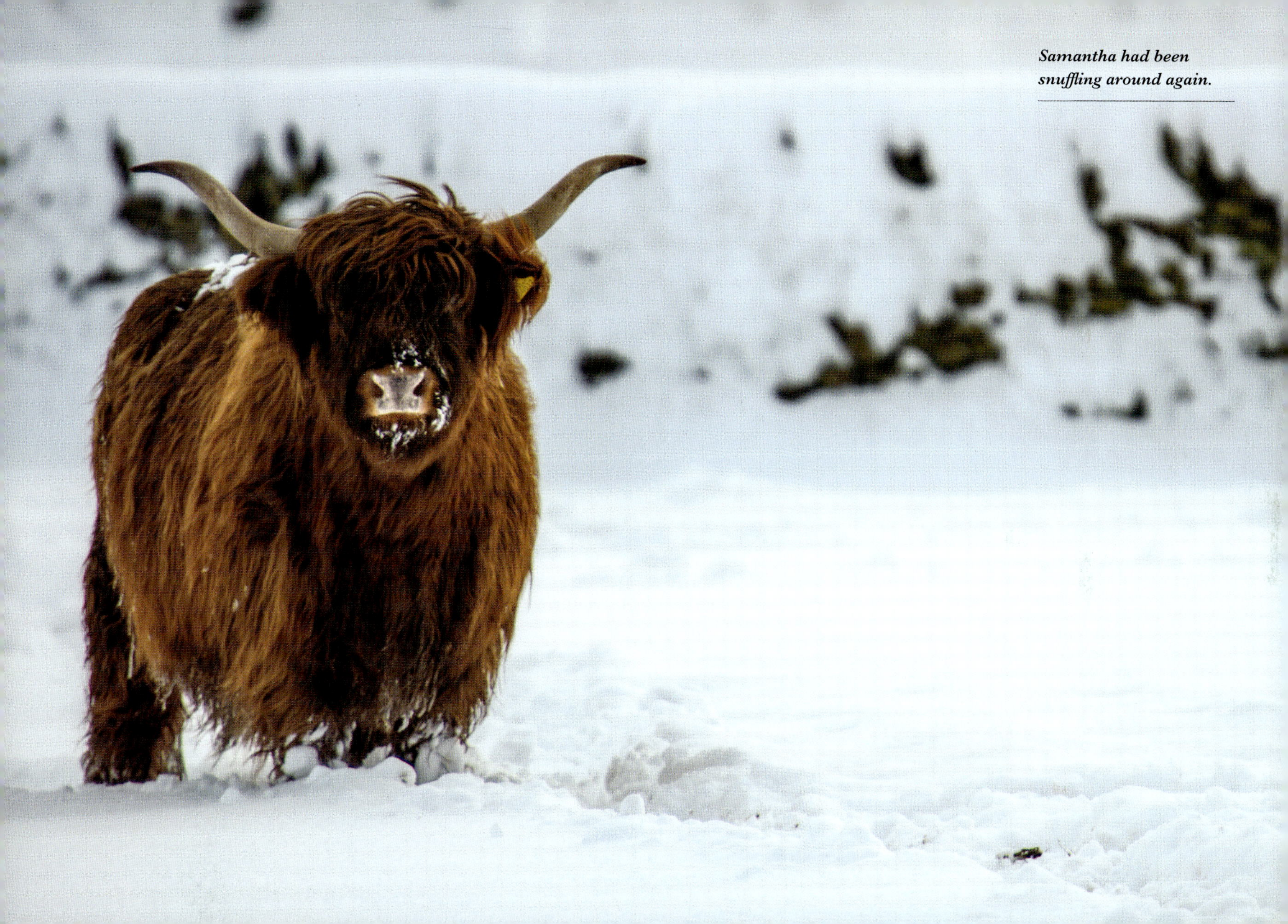

Samantha had been
snuffling around again.

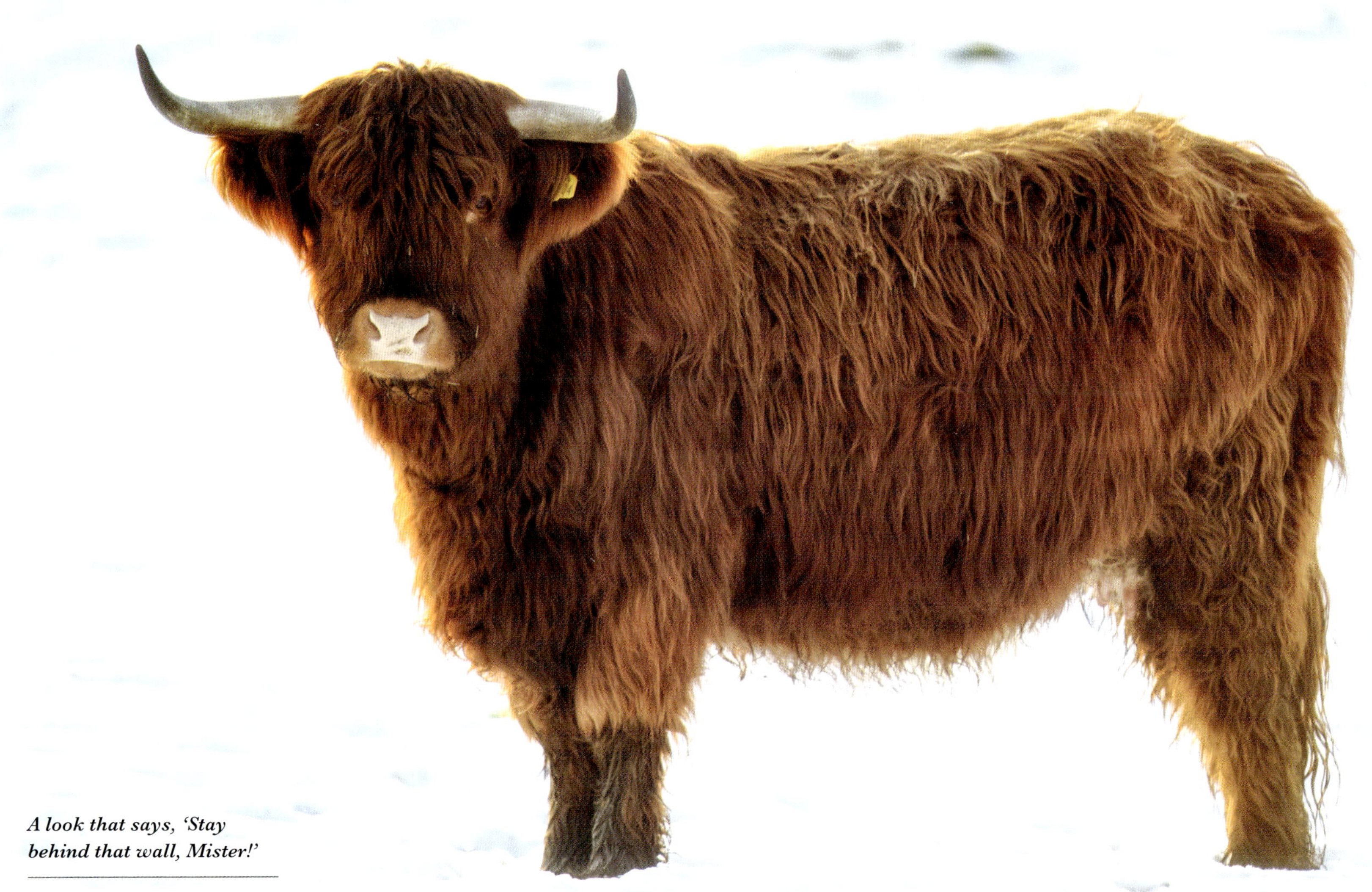

*A look that says, 'Stay
behind that wall, Mister!'*

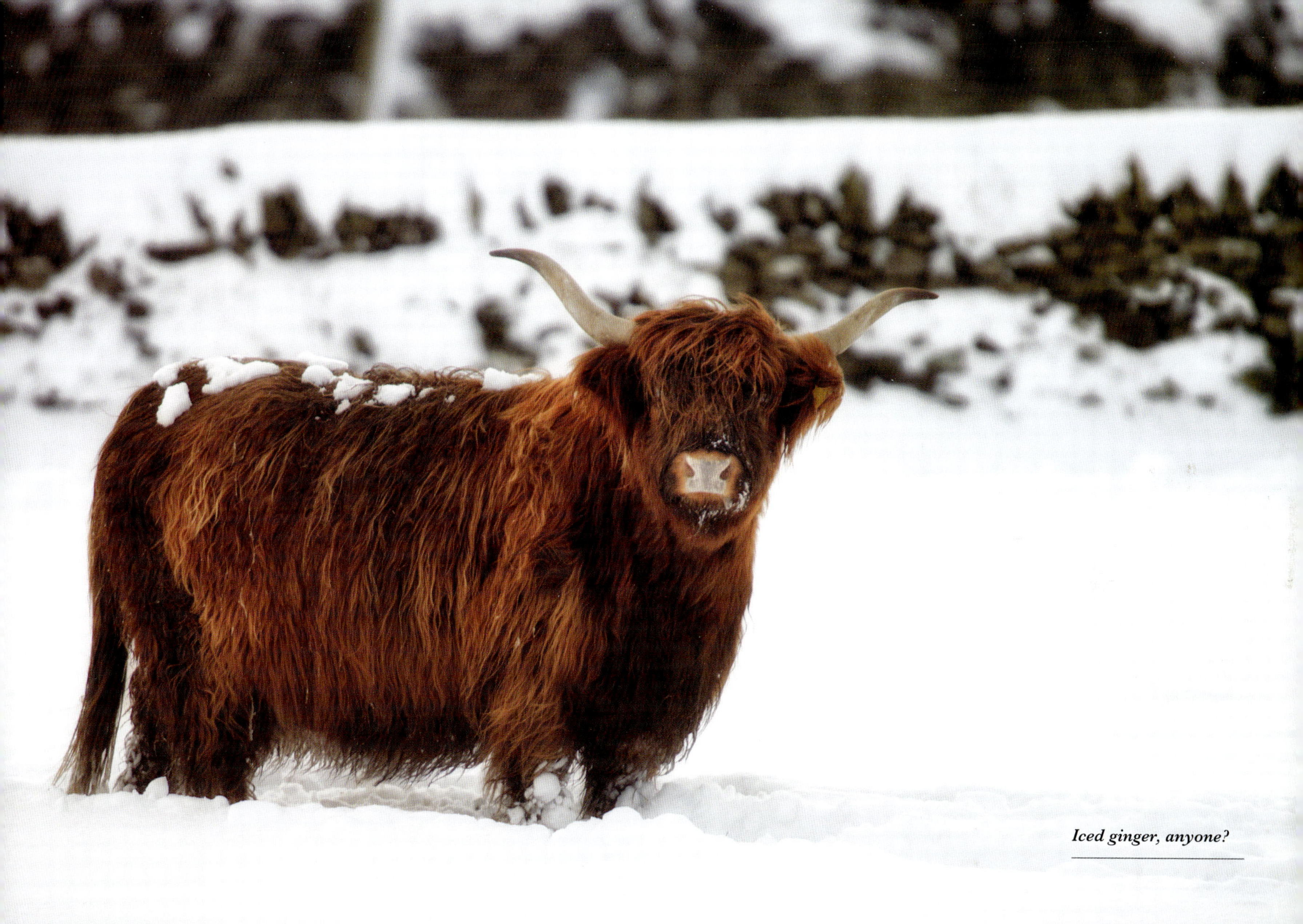

Iced ginger, anyone?
____

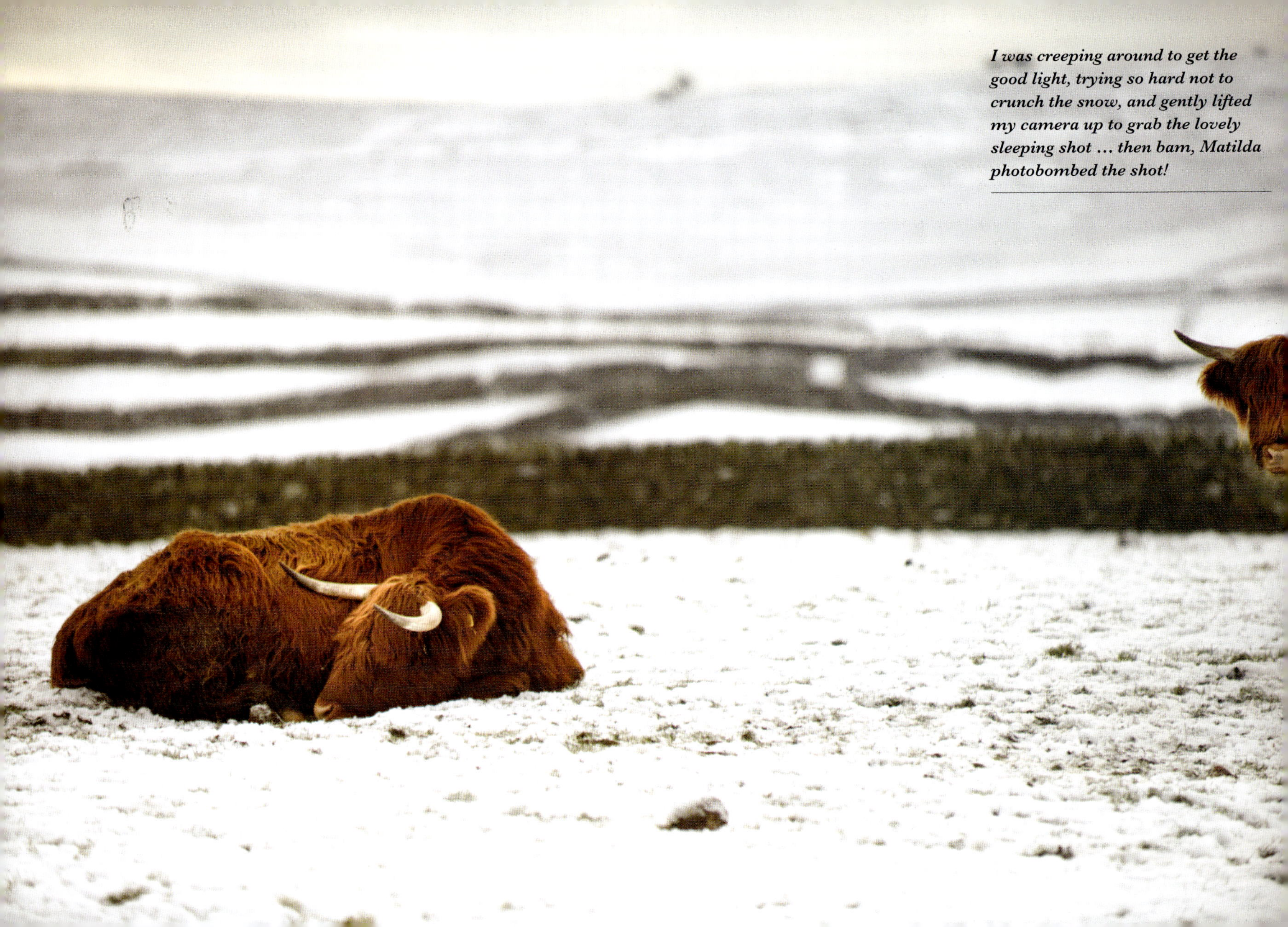

I was creeping around to get the
good light, trying so hard not to
crunch the snow, and gently lifted
my camera up to grab the lovely
sleeping shot ... then bam, Matilda
photobombed the shot!

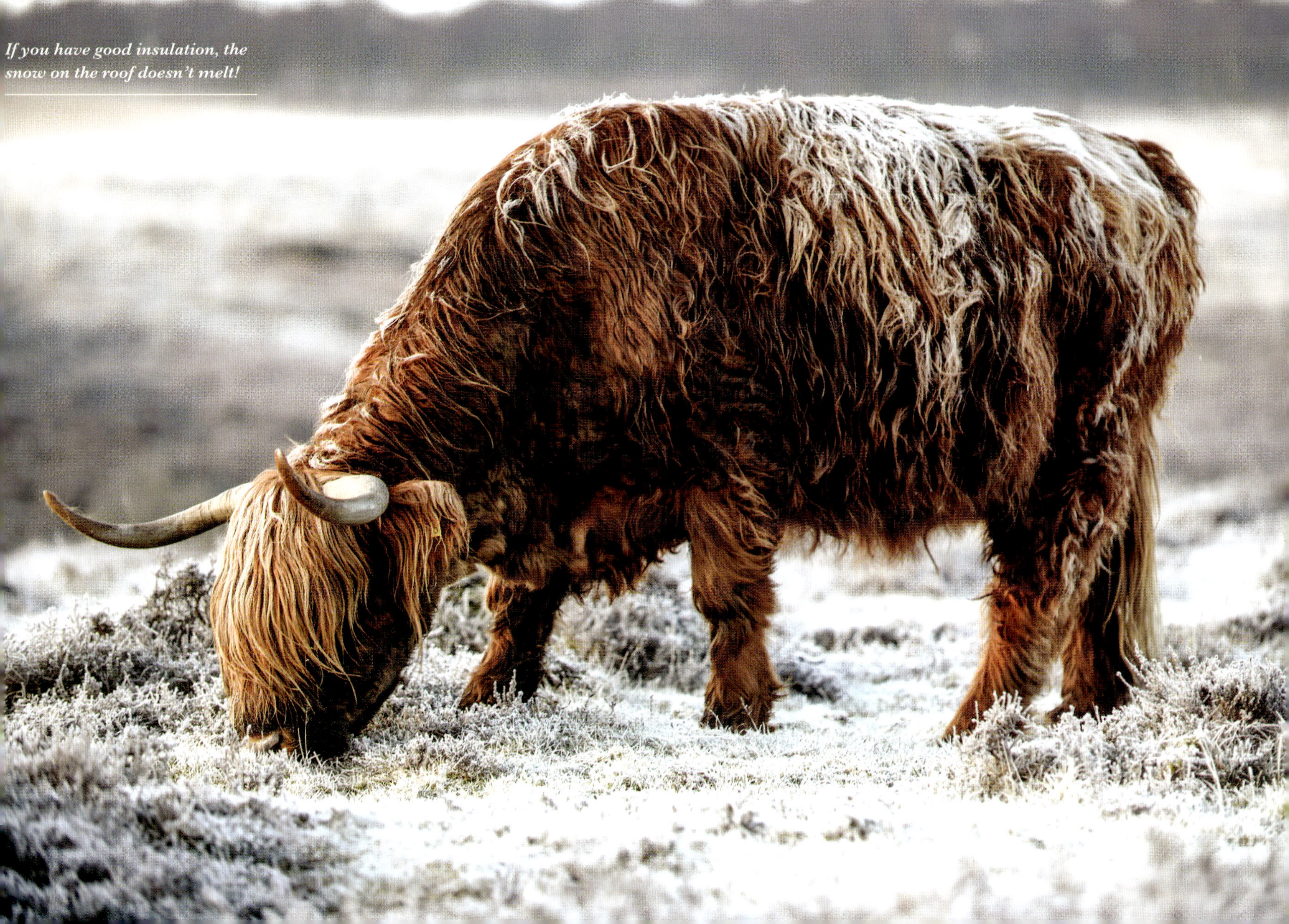
If you have good insulation, the snow on the roof doesn't melt!

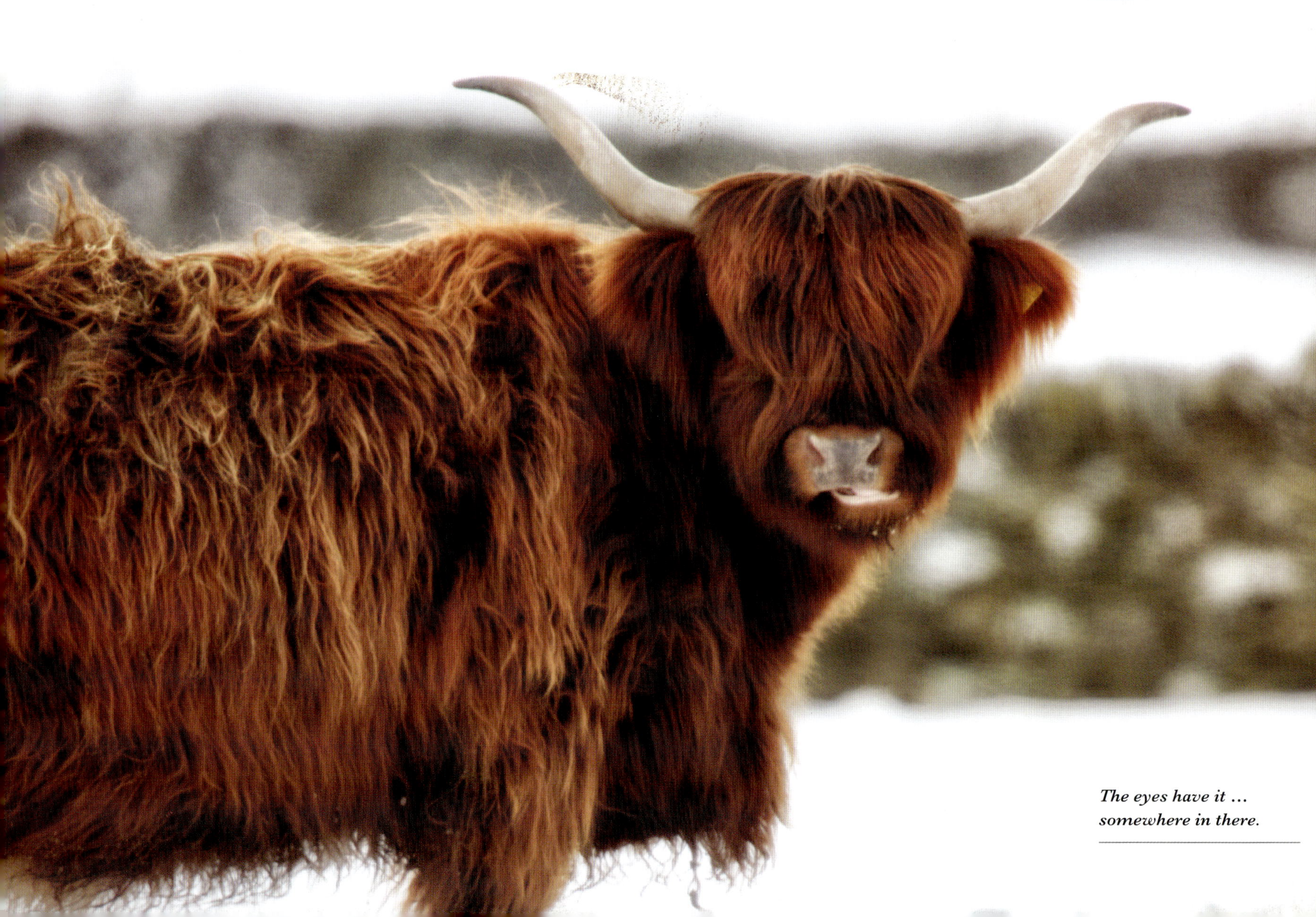

*The eyes have it …
somewhere in there.*

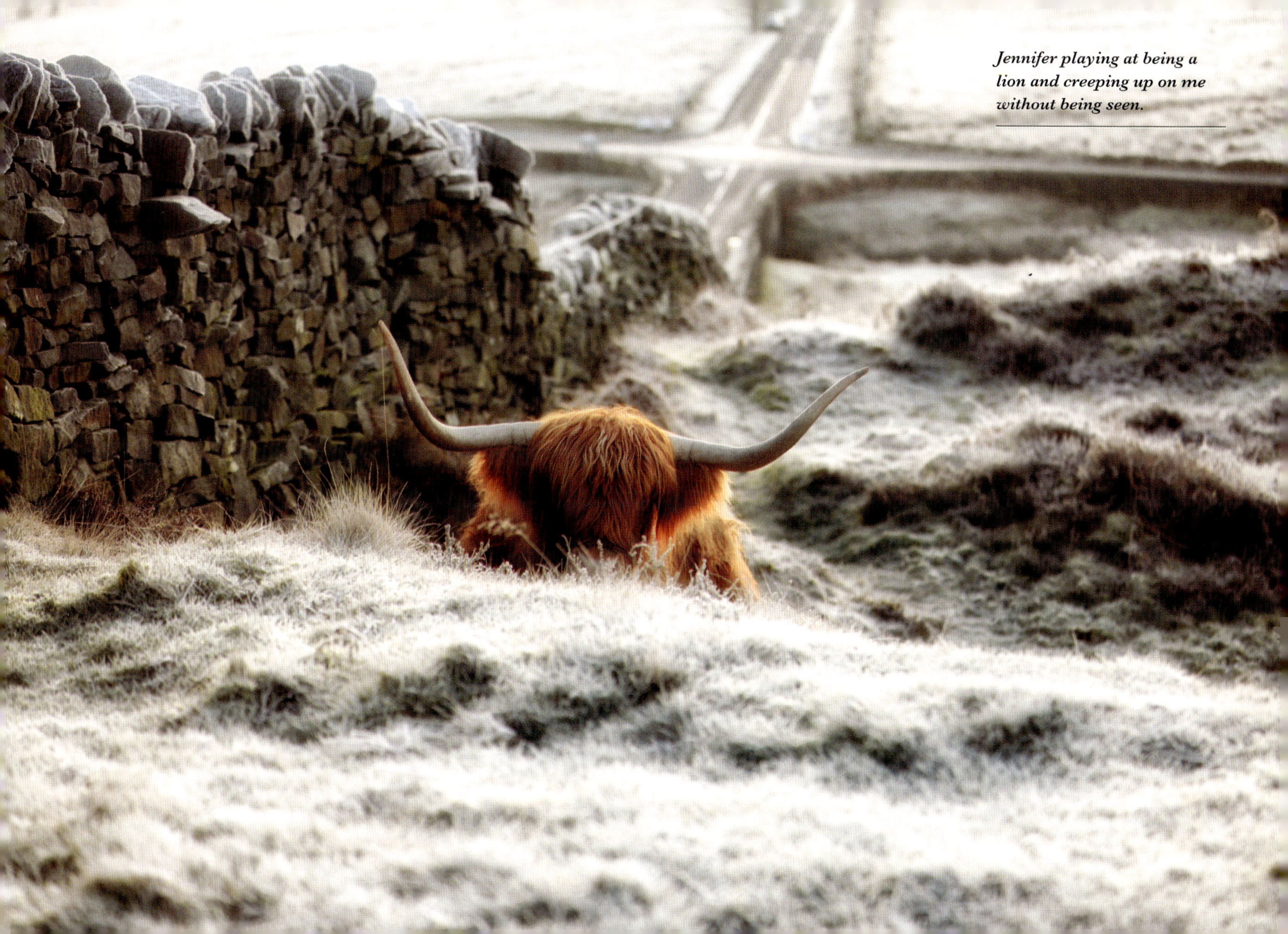

Jennifer playing at being a
lion and creeping up on me
without being seen.

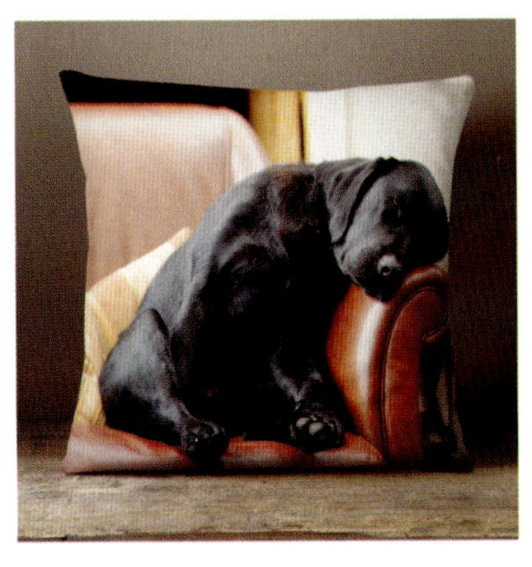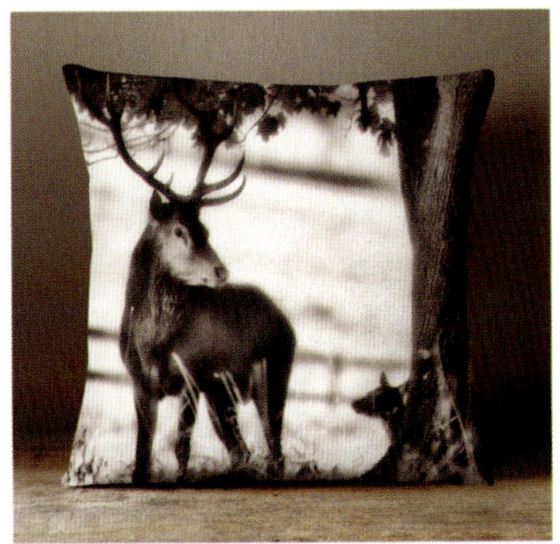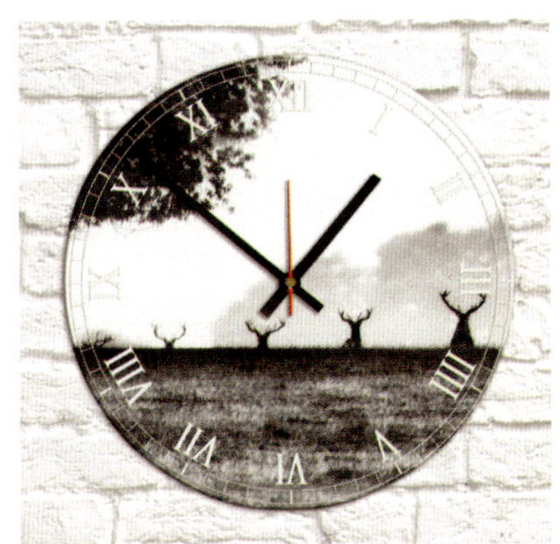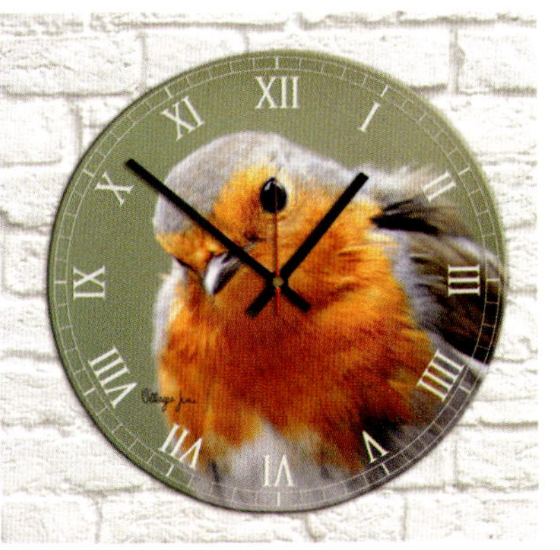